U0715367

曹福亮 (Fuliang Sam Cao) 主编

中国银杏品种图鉴

An Illustrated Monograph of *Ginkgo biloba* L. Cultivars in China

科学出版社

主　编	曹福亮					
副主编	张往祥	朱灿灿	赵洪亮			
编　委	(按姓氏笔画排序)					
	马有基	邓荫伟	史锋厚	田亚玲	李　群	吴家胜　苏明洲
	汪贵斌	周吉林	欧祖兰	罗文金	郁万文	宫玉臣　赵　志
	郝明灼	熊　壮	潘小平			

图书在版编目（CIP）数据

中国银杏品种图鉴/曹福亮主编．—北京：科学出版社，2010
ISBN 978-7-03-026858-7

I.中… II.①曹… III.银杏－品种－中国－国集 IV.S664.302.4-64

中国版本图书馆CIP数据核字（2010）第032718号

策划编辑：李　锋　李振格／责任编辑：童安齐　田斩峰
责任校对：马英菊／责任印制：吕春珉
装帧设计：曹　来／制版：北京美光制版有限公司

科学出版社 出版
北京东黄城根北街16号
邮政编码：100717
http://www.sciencep.com

北京华联印刷有限公司　印刷
科学出版社发行　各地新华书店经销

*

2011年4月第　一　版　　开本：210×285
2011年4月第一次印刷　　印张：15 ½
印数：1—1 500　　　　　字数：350 000

定价：160.00元
（如有印装质量问题，我社负责调换）

主编简介
About the Author

曹福亮（英文名 Sam）先后获南京林业大学学士（林学专业）学位，南京林业大学农学硕士（森林培育学专业）学位，加拿大不列颠哥伦比亚大学哲学博士（森林生态学专业）学位。现任南京林业大学校长，中国林学会银杏分会（原中国银杏研究会）主任委员，中国银杏产业联谊会会长，中国林学会经济林分会副主任委员，中国林学会森林培育学分会副主任委员，中国园艺学会干果分会副主任委员，国家级林学实验教学中心主任，江苏省"333"中青年首席科学家，江苏省特种经济树种培育与利用工程技术研究中心主任。

长期以来，主要从事经济林栽培及经济植物资源开发利用等方面的教学和科研工作。近年来，以银杏研究为特色，在银杏分子生物学、遗传和进化、良种选育、抗性、种间和种内竞争、培育机理和综合开发利用等方面开展了全面和系统的研究，主持 20 多项国家和省部级攻关课题，获得国家级和省部级科技奖励 8 项，其中，"银杏、杨树、落羽杉等三个树种抗性机理研究"和"银杏等四个树种良种选育及培育技术研究与推广"分别于 2003 年和 2007 年获国家科技进步二等奖，"银杏资源综合加工利用"获 2007 年度梁希科技进步一等奖。

先后在国内外学术刊物上发表论文 160 余篇，出版《中国银杏》、《中国银杏志》、《银杏》（画册）、《银杏资源培育与高效利用》和 *Forest Ecology* 等著作。

Fuliang Sam Cao Following an undergraduate degree in Forestry Sciences and an MSc in Silviculture from Nanjing Forestry University (NJFU), Nanjing, P.R.China, Fuliang Sam Cao received his PhD in Forest Ecology at the University of British Columbia, Canada, focusing on Ecological Basis for Ginkgo Agroforestry. Since 1982 he has been working as one of the faculty member at NJFU, where he presented courses on silviculture, no-wood products, bamboo forests, methodology for no-wood products research.

For the past twenty years, Dr Cao has undertaken many key research projects mainly related to the ecophysiology, forest ecology and silviculture of poplar, bamboo and ginkgo. He has published several books titled as *Chinese Ginkgo, Ginkgo, Growth Dynamics of Southern Poplar Clones*, and *Forest Ecology*. Since 1992, his research interests have been focused on ginkgo tree species covering many aspects from genetics to physiology, forest ecology, silviculture and utilizations of ginkgo. The forty-eight graduate students he has supervised over the past ten years have studied all aspects of Silviculture and some areas of Forest Ecology covering ecophysiology, production ecology, agroforestry, tree species selection, density control and modeling.

Dr Cao is currently professor of Silviculture and President of NJFU and Chairman of Chinese Ginkgo Research Institute.

银杏，我喜欢你，
你是真应该成为中国的国树呀，
我是喜欢你，
我特别的喜欢你。

——《银杏》·郭沫若

序

本人是观赏园艺的老兵,对古树名木情有独钟。对于与恐龙同时代的"活化石"银杏,更持有无限尊重和关爱之情。我年轻时学习和工作在成都,就曾几次到灌县青城山天师洞,观看并记载那栽于院旁的冠幅达36米的汉代银杏。八旬之后,又专赴山东莒县浮来山定林寺,一览树龄达三四千年、世界第一古银杏的风采。前几年结识南京林业大学曹福亮教授,他研究银杏多年,成绩斐然。近年国人评选国树,银杏的呼声最高,曹教授在其书中亦表赞同。本人也认为以银杏为我国国树,当系众望所归,是国内外一致拥护的。

近读曹教授几部专著,深感他既重学术专著,又不忘科普宣传。这种兼筹并顾的做法,必可让银杏扬名于天下,让大家都了解银杏、宣传银杏,使银杏更好地造福于中国,并惠及世界人民。

曹教授既从古植物历史上研究银杏,又从世界角度来探讨它,故其著述中所介绍的银杏,就更显得鹤立鸡群,入木三分。

著者善于与时俱进,不断扩大研究与应用的领域。如对银杏,除了观赏和木材两项固有用途外,还对国内多处种核产地及品种进行了系统调查与研究,近年还开展了银杏叶药用品种调查与分类的探讨。《中国银杏品种图鉴》记载了各类银杏品种和无性系100多个,这是著者多年研究银杏的最新总结,其研究成果确为国人和世界做出了全方位的贡献。

在曹教授专著及其他书刊中,都孕育着银杏科学与银杏文化的交融,这是中华文明的特色之一,是应予弘扬光大的。

《中国银杏品种图鉴》即将出版,这是我国申报银杏品种国际登录权威的重要著作和依据。我对该书出版喜讯表示祝贺,并略陈曹福亮教授其人其书之若干特色,以飨读者,是为序。

九二叟 陈俊愉
于 北京林业大学
2009年1月28日
农历正月初三

左图:南京鼓楼广场银杏景观
Ginkgo in Gulou Square, Nanjing

前 言

银杏(*Ginkgo biloba* L.)，是银杏目植物中现存的唯一种。经过长期的栽培选育，银杏逐渐演化、发展形成了现在极为丰富的银杏品种群，其中具有悠久栽培历史的中国及具有较长引种史的日本有着丰富的核用品种，而美国和中国则有着丰富的观赏品种。

由于银杏在世界的分布范围十分广泛，加上选育目的不同和国际间交流较少等原因，目前的银杏品种命名和分类存在以下三个方面的问题：一是银杏虽具有多方面的用途，但一提到品种则多指银杏的核用品种，而对叶用品种、材用品种、观赏品种等涉及很少；二是各银杏产区多年来选育出的传统优良品种（如'泰兴佛指'、'洞庭佛手'、'郯城金坠'、'马铃'、'郯城圆铃'等）经过长期的栽培出现了很多变异，这些变异性状有的已超过原有品种的品质，因而在生产实践中又选育出许多新"优良品种"（如'郯城5号'、'郯城9号'、'圆铃13号'等），由于分类依据和方法的不同，造成品种名称的混乱；三是由于群众迫切需要银杏良种，而银杏良种又均系嫁接植株，很容易造成一种错误印象，凡是嫁接银杏一律都是优良品种，这给商品交易中的假冒伪劣品种以可乘之机，造成银杏品种的很大混乱。另外，我们在整理银杏品种的过程中，发现许多品种是同物异名，如'家佛指'、'泰兴大白果'、'大佛指'为同一品种；有些是同名异物，如'大金果'（又名'魁铃'）和'大金果'（'铁富4号'）分别是郯城和邳州选育的两个不同品种；还有一些品种名称是栽培者或选育者临时取名但沿袭至今。在银杏主产区，普遍存在着品种混合栽植的现象，我们在整理银杏品种的过程中，尽可能地规范统一。鉴于此，为了更好地对银杏品种进行分类和鉴别，在生产中推广优良品种，有必要在规范统一的原则（《国际栽培植物命名法规》）下，对中国及国外的银杏品种进行汇总，科学地分类和命名。

多年来，南京林业大学先后承担了多项国家、部（省）级重点课题和横向银杏课题，重点开展以银杏为特色的经济树种培育和加工理论技术研究，特别是在银杏种质基因库建立、遗传进化分析以及杂交育种、良种选育等方面开展了系统的研究。在江苏江都、泰兴、下蜀等地建成了我国拥有银杏种质资源最全、最多的银杏基因库，开展了银杏资源的遗传多样性分析，并开展了银杏分子进化研究，研制了一张密度较高的银杏遗传图谱，率先开展以核用、观赏用、叶用、花粉用、外种皮用为目的的遗传育种工作。这些研究成果以及国内外其他一些学者在银杏品种分类方面的研究成果，为本书的出版提供了良好的素材和基础。

全书以中英文双语形式全面系统地介绍了银杏栽培品种的生态生物学习性、繁殖、栽培和应用，阐述了银杏品种群和品种分类的原则和方法，编制了含43

个银杏核用品种、13个观赏品种的品种分类系统检索表，其中核用品种根据银杏种核形态重要指标，分为四大品种群，即长子品种群、佛指品种群、中子品种群和圆子品种群。在此基础上，以标本和照片的形式对四大类群中的一些重要品种及优良无性系进行重点介绍。在编写过程中，我们还清理了许多同名异物或异名同物的品种，并淘汰了部分"无价值"的品种，严格按照最新《国际栽培植物命名法规》逐一进行整理，力求每一个记载品种的准确性、典型性和科学性。

在本书撰写过程中得到了各界领导和同仁的大力支持和无私帮助。梅花国际登录权威陈俊愉院士在百忙中为本书作序，木犀属品种国际登录权威向其柏教授为本书审稿，并为该书检索表的制作和品种描述提供了许多建设性的指导意见，北京林业大学张志成教授和南京林业大学黄鹏成教授对该书品种检索表的研制提出了修改意见，山东省郯城县林业局宫玉臣副局长和苏明洲主任、江苏省邳州市多管局赵洪亮副局长和邳州银杏研究所孟宪峰所长、江苏省泰兴市林业局李群副局长、广西师范大学邓荫伟教授、湖北省安陆市林业局罗文金局长和潘小平副局长、浙江省天目山自然保护区赵明水总工程师和浙江省临安市昌化林业分站周荣高工程师等对银杏品种的收集提供了大量的帮助和支持，南京林业大学摄影中心马有基教授和研究生赵志承担了书中大多数图片的拍摄任务，南京林业大学树木学组袁发银老师指导了标本的采集和制作。另外，在本书的编写、资料整理、文字翻译、标本制作、图片处理等工作中，湛磊老师和郁万文、熊壮、田亚玲、欧祖兰、蔡金峰、陈军、Pius. A. Adeniyi 等多位研究生也付出了辛勤的劳动，科学出版社的童安齐、田靳峰等同志为该书的出版倾注了大量的心血，在此一并表示谢意。

本书力求科学、系统、全面地反映当今银杏种质资源状况及品种分类研究的最新成果，希望该书的出版能为中国银杏业的发展做一点有益的贡献。但由于笔者水平有限，难免有不足之处，敬请读者批评指正。

曹福亮
2010 年 4 月

目 录

序

前言

第1章 银杏的形态特征 ················ 1

第2章 银杏品种分类依据 ················ 11

第3章 银杏品种检索表及品种图谱 ················ 19

 3.1 银杏核用品种及无性系 ················ 20

  3.1.1 长子品种群 ················ 22

   3.1.1.1 长子品种群分类检索表 ················ 22

   3.1.1.2 长子品种群优良品种 ················ 24

   3.1.1.3 长子品种群优良无性系 ················ 34

  3.1.2 佛指品种群 ················ 50

   3.1.2.1 佛指品种群分类检索表 ················ 50

   3.1.2.2 佛指品种群优良品种 ················ 52

   3.1.2.3 佛指品种群优良无性系 ················ 78

  3.1.3 中子品种群 ················ 105

   3.1.3.1 中子品种群分类检索表 ················ 106

   3.1.3.2 中子品种群优良品种 ················ 108

   3.1.3.3 中子品种群优良无性系 ················ 136

  3.1.4 圆子品种群 ················ 151

   3.1.4.1 圆子品种群分类检索表 ················ 152

   3.1.4.2 圆子品种群优良品种 ················ 154

   3.1.4.3 圆子品种群优良无性系 ················ 176

 3.2 银杏观赏品种 ················ 183

  3.2.1 银杏观赏品种分类检索表 ················ 184

  3.2.2 银杏观赏品种描述 ················ 186

 3.3 银杏叶用优良无性系 ················ 196

 3.4 银杏花粉用优良无性系 ················ 206

 3.5 银杏材用优良无性系 ················ 210

第4章 银杏古树资源 ················ 213

参考文献 ················ 233

索引 ················ 240

Contents

Preface
Foreword
Chapter 1　Morphological Characteristics of *Ginkgo biloba* L. ··· 1
Chapter 2　Classification Basis of *Ginkgo Cultivars* ············11
Chapter 3　Key to Cultivars of *Ginkgo biloba* L. and Illustration of
　　　　　　Ginkgo Cultivars ·· 19
　　3.1　Ginkgo Cultivars and Clones for Nut-producing ···················· 21
　　　　3.1.1　Changzi Group ··· 23
　　　　　　3.1.1.1　Key to Changzi Group ································· 23
　　　　　　3.1.1.2　Cultivars of Changzi Group ··························· 24
　　　　　　3.1.1.3　Clones of Changzi Group ······························ 34
　　　　3.1.2　Fozhi Group ·· 51
　　　　　　3.1.2.1　Key to Fozhi Group ··································· 51
　　　　　　3.1.2.2　Cultivars of Fozhi Group ····························· 52
　　　　　　3.1.2.3　Clones of Fozhi Group ································ 78
　　　　3.1.3　Zhongzi Group ··· 105
　　　　　　3.1.3.1　Key to Zhongzi Group ································ 107
　　　　　　3.1.3.2　Cultivars of Zhongzi Group ··························· 108
　　　　　　3.1.3.3　Clones of Zhongzi Group ····························· 136
　　　　3.1.4　Yuanzi Group ·· 151
　　　　　　3.1.4.1　Key to Yuanzi Group ································· 153
　　　　　　3.1.4.2　Cultivars of Yuanzi Group ··························· 154
　　　　　　3.1.4.3　Clones of Yuanzi Group ······························ 176
　　3.2　Ginkgo Cultivars for Ornamental Purpose ·························· 183
　　　　3.2.1　Key to Ginkgo Cultivars for Ornamental Purpose ········· 185
　　　　3.2.2　Illustration of Ornamental Cultivars ····················· 186
　　3.3　Ginkgo Clones for Leaf-producing ································ 197
　　3.4　Ginkgo Clones for Pollen-producing ······························ 206
　　3.5　Ginkgo Clones for Timber-producing ····························· 211
Chapter 4　Resources of Ancient Ginkgo Trees in China ········· 213
References ·· 233
Index ··· 240

浙江天目山"五代同堂"
Ginkgo Tree with Different Ages Main Stems in Tianmu Mountain, Zhejiang

第 1 章 Chapter 1

银杏的形态特征

Morphological Characteristics of Ginkgo biloba L.

第1章 银杏的形态特征
Chapter 1　Morphological Characteristics of *Ginkgo biloba* L.

银杏(*Ginkgo biloba* L.)也称白果树,是银杏科(Ginkgoaceae)、银杏属(*Ginkgo*)的单属种植物,雌雄异株。银杏起源于距今3亿多年前的古生代石炭纪,经过第四纪冰川后仅在中国长江流域一带的群山中保存了下来,因而有植物活化石之称,是现存裸子植物中与恐龙同时代的最古老的孑遗植物。银杏,融食用、药用、保健和生态等功能于一体,集自然景观和人文景观于一身,银杏产业已开发利用发展成为当今世界规模化、多元化、集约化、标准化的国际性高效产业,产生了十分重大的经济效益、社会效益、生态效益和景观效益,正在为人类作出越来越大的贡献。

银杏抗病虫害能力特别强,适应范围广,银杏耐旱、耐寒、耐高温、耐瘠薄、耐核辐射能力强,银杏浑身都是宝,因此银杏是重要的经济生态型树种。

银杏大多是高大乔木,幼树树皮浅纵裂,大树的皮呈灰褐色,深纵裂,粗糙;枝近轮生,斜上伸展(雌株的大枝常较雄株开展);1年生的长枝淡褐黄色,2年生以上变成灰色,并有细纵裂纹;短枝密被叶痕,黑灰色,短枝上亦可长出长枝;冬芽黄褐色,常为卵圆形,先端钝尖(图1.1,图1.2,图1.5)。

Ginkgo biloba L., Maidenhair tree, is a "living fossil", being the only surviving representative of the Ginkgoales (an order placed between the conifers and the cycads) which were widespread in Jurassic and Cretaceous times of 195-65 million years ago. *Ginkgo biloba* and other species of the genus were once widespread throughout the world in history. In the end of tertiary and the initial stage of Quatemary, ginkgo was destroyed in many regions, surviving only in the Yangtze river basin of central China where the modern species survived. *Ginkgo biloba* is now a rare species in wild, but has been widely cultivated as an ornamental, probably for more than 3 000 year. Wood is used in furniture making; leaves are medicinal and used for pesticides, roots are used as a cure for leucorrhea, fruits are edible, and bark yields tannin. The easily worked wood is valued and used in joinery. Leaves of Ginkgo are rich in special chemical compounds such as flavone, biflavone, flavone glycoside, and Ginkgolide, which can be used in medicines for cardiac vessel and hypertension diseases, and cosmetics additives and hair-promoting agents.

Ginkgo has strong diseases and pests resistant, grows in any place. Ginkgo has many good characters, such as multi-resistance,drought, low temperature,barren, and nuclear radiation. The wholesale tree of gingko is valuable. It is the important economical ecological tree speices.

Ginkgo trees are very tall, bark light gray or grayish brown, longitudinally fissured especially on old trees normally; crown conical initially, finally broadly ovoid; branches are nearly whorled, assurgent and extended; branches pale brownish yellow initially, finally gray; dwarf branches blackish gray, with dense, irregularly elliptic leaf scars; winter buds yellowish brown, ovate (Fig. 1.1, Fig. 1.2, Fig. 1.5).

图1.1 银杏形态图（蒋杏强绘）
1.雌球花枝；2.雌球花上端；3.长短枝及种实；4.去外种皮的种子；5.去外、中种皮的种子纵切面（示胚乳与子叶）；6.雄球花序；7.雄蕊

Fig.1.1　Morphological Graph of Ginkgo (By Jiang Xingqiang)
1. Ovules cones, 2. Top of ovules cones, 3. Long and dwarf branches with fruits, 4. Nut, 5. Longitudinal section of kernel (endosperm and cotyledon), 6. Pollen cones, 7. Stamen

第 1 章 银杏的形态特征
Chapter 1 Morphological Characteristics of *Ginkgo biloba* L.

银杏的树冠，因年龄、性别、品种和繁殖手段的不同而呈各种姿态（图1.4）。银杏的枝条分长枝和短枝。长枝是指从主干上生长出来的骨干枝、各级骨干枝上长出来的下垂母枝、下垂母枝上长出来的下垂枝。短枝是由长枝中下部的腋芽所形成，花、果均着生在短枝上，与叶混生，呈螺旋状排列。银杏短枝中只有一个顶芽，外被鳞片，呈覆瓦状，发芽后鳞片脱落，根据短枝鳞片脱落的痕迹，可清楚数出短枝的年龄（图1.5）。

The forms of Ginkgo crown are various in different ages, gender, varieties and breeding practice (Fig. 1.4). There are two types of branches, long branches and dwarf branches. Dwarf branches dimorphic: both long and short. There are many patterns of long branches; diaphysis branches, pendulous branches from different levels diaphysis branches, pendulous branches from pendulous branches. Dwarf branches are developed from auxiliary buds in the lower part of the long branches. Flowers, fruits are spirally arranged on dwarf branches, mixed with leaves. There is only one terminal bud, which tile-like covered by scales (Fig. 1.5).

图 1.2
银杏主干和树皮形态
① 通直的主干；
② 树瘤；
③ 深裂状树皮；
④ 鳞片状树皮

Fig. 1.2
Stems and Barks of Ginkgo
① Straight stem,
② Tree wart,
③ Deep-fissured bark,
④ Scaly bark

银杏的形态特征
Morphological Characteristics of *Ginkgo biloba* L.

图 1.3 银杏树奶

Fig. 1.3 The Chichi

图 1.4 银杏树冠形态
①圆锥型树冠；②塔形树冠；③圆柱形树冠；④垂枝形树冠

Fig. 1.4 Different Crowns of Ginkgo
① Conical crowns, ② Pyramid crowns, ③ Cylindrical crowns, ④ Weeping crowns

第1章 银杏的形态特征
Chapter 1 Morphological Characteristics of *Ginkgo biloba* L.

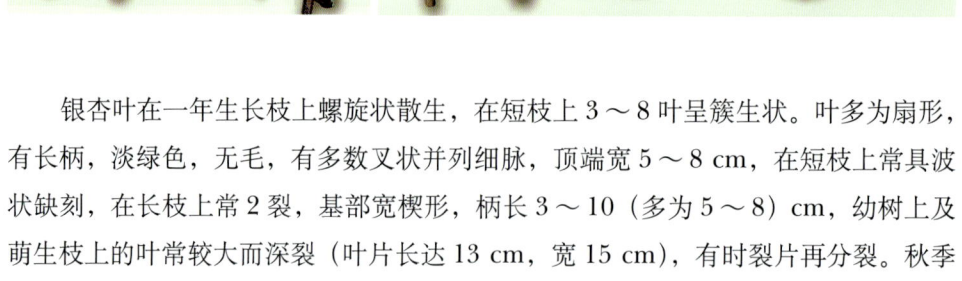

图 1.5
银杏芽和枝条

**Fig. 1.5
Buds and Branches of Ginkgo**

　　银杏叶在一年生长枝上螺旋状散生，在短枝上 3~8 叶呈簇生状。叶多为扇形，有长柄，淡绿色，无毛，有多数叉状并列细脉，顶端宽 5~8 cm，在短枝上常具波状缺刻，在长枝上常 2 裂，基部宽楔形，柄长 3~10（多为 5~8）cm，幼树上及萌生枝上的叶常较大而深裂（叶片长达 13 cm，宽 15 cm），有时裂片再分裂。秋季落叶前变为黄色，有的银杏品种叶片具黄色条纹（图 1.6）。

　　Leaves sparsely and spirally arranged on long branches, fasciculate on dwarf branches, long petiolate, flabellate, venation parallel, close, dichotomous, open but with rare anastomoses, base broadly cuneate, margin straight, entire, apex 2-lobed or notched. Leaves with petiole 3–10 cm; blade pale green, to 13 × 8 (–15) cm on young trees but usually 5–8 cm wide, those on long branches divided by a deep, apical sinus into 2 lobes each further dissected, those on dwarf branches with undulate distal and margin notched apex. Leaves turning bright yellow in autumn, some species have yellow stripe (Fig. 1.6).

图1.6
银杏叶

Fig. 1.6
Leaves of Ginkgo

第1章 银杏的形态特征
Chapter 1 Morphological Characteristics of *Ginkgo biloba* L.

银杏球花雌雄异株，单性，生于短枝顶端的鳞片状叶的腋内，呈簇生状；雄球花柔荑花序状，下垂，雄蕊排列疏松，具短梗，花药常2个，长椭圆形，药室纵裂，药隔不发达；雌球花具长梗，梗端常分两叉，稀3～5叉或不分叉，每叉顶生一盘状珠座，胚珠着生其上，通常仅一个叉端的胚珠发育成种子（图1.7和图1.8）。

种实具长梗，下垂，常为椭圆形、长倒卵形、卵圆形或近圆球形，外种皮肉质，熟时黄色或橙黄色，外被白粉，有臭味；中种皮白色，骨质，具2～3纵脊；内种皮膜质，淡红褐色；胚乳肉质，子叶2枚，稀3枚，发芽时不出土，花期3～4月，种子9～10月成熟（图1.9）。

Reproductive structures produced in clusters in axils of scalelike leaves at apex of dwarf branches before leaves expand. Pollen cones pedunculate, pendulous, catkinlike; microsporophylls numerous, spirally and rather sparsely arranged; microsporangia 2, elliptic; pollen sacs longitudinally dehiscent. Ovules borne on a long, dichotomously branched peduncle sometimes dividing into 3–5 terminal forks; each fork discoid at apex, bearing an erect, sessile ovule; usually only 1 fruits ripening per peduncle (Fig. 1.7 and Fig. 1.8).

Fruits long pedunculate, pendulous, drupelike, with a single integument differentiated into a fleshy, succulent sarcotesta, yellow, or orange-yellow, with rancid odor when ripe; a bony sclerotesta, white, with 2 or 3 longitudinal ridges;and a membranous endotesta, pale reddish brown; female gametophyte tissue abundant. Gametophyte tissue abundant. Cotyledons 2 or 3. Germination hypogeal. Pollination Mar.–Apr., fruits maturity from Sep. to Oct. (Fig. 1.9).

图 1.7
银杏雌球花和雄球花
①雌球花；
②雌球花授粉滴；
③雄球花

Fig. 1.7
Ovules and pollen cones of Ginkgo
① Ovules,
② Pollination droplet of Ginkgo ovules,
③ Pollen cones

银杏的形态特征
Morphological Characteristics of *Ginkgo biloba* L.

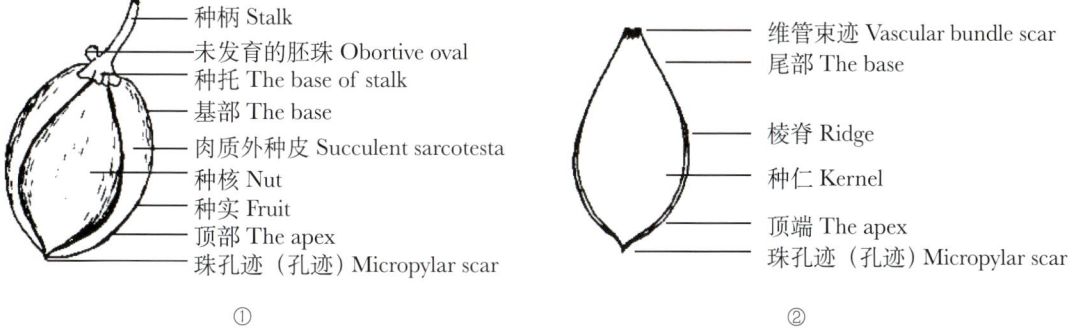

图 1.8　银杏种实和种核结构简图
①银杏种实；②银杏种核
Fig. 1.8　Structure Graphs of Ginkgo Fruit and Nut
① Fruit, ② Nut

图 1.9
银杏种实
①种实；
②叶籽银杏；
③未成熟果；
④成熟果

Fig. 1.9
Fruits of Ginkgo
① Fruit,
② Ye Zi Yinxing
③ Immature fruits,
④ Mature fruits

硕果累累
Full of Fruits

第2章 Chapter 2

银杏品种分类依据

Classification Basis of Ginkgo Cultivars

第2章 银杏品种分类依据
Chapter 2　Classification Basis of *Ginkgo Cultivars*

银杏（*Ginkgo biloba* L.）经过3亿年以上的沧桑巨变，由发生到葳蕤葱茏，由种类繁多到第四纪冰川浩劫后成为仅存我国栽培的单科属种，是世界上最古老的孑遗植物，被公认为"活化石"。尽管银杏是单科属种，但由于银杏雌雄异株及其分布的广泛性和栽培目的的多样性，使其在长期的系统发育和个体发育中产生了较多的变异，从而形成了银杏不同类型的丰富资源。

银杏作为单科属种植物，在植物分类中仅为很小的一群。由于长时间的栽培，通过天然杂交和人工选择，银杏种子和叶片均发生了变异，加之地区之间的隔离，产生了较多的银杏品种（品系、优株、无性系），但在分类上没有统一的标准，有待于进一步研究，并加以规范。

1. 银杏品种分类

1771年，林奈（Linnaeus）根据德国医生肯普弗（Kaempfer）在日本所采银杏标本及其形态描述，正式命名了银杏的学名为 *Ginkgo biloba* Linn.。自此之后，各国学者对银杏种级以下的分类进行了一个多世纪的探索。不同学者依照各自的观察标准和分类目的，采取了不同的分类方式、不同的分类等级和不同的分类命名方法。

E. A. Caarriere 于1854年根据叶片的形态发表了两个分类单位，即 *Salisburia adiantifolia laciniata* Carr. 裂叶银杏（大叶银杏）和 *Salisburia adiantifolia variegata* Carr. 斑叶银杏（花叶银杏）。他把银杏种级以下都定为变型，这就奠定了以后银杏分类的基础。后来的一个多世

Ginkgo biloba L. has survived for 3 hundred-million years, almost all the species of ginkgo were extinct except one in China. It is one of the most ancient relic plants in the world, it has been widely received as a living fossil. Ginkgo is dioecious, widely distributed in the world. Ginkgo is rich in resources because of ginkgo has many varieties not only ginkgo is dioecious but also ginkgo is widely distributed and has multiaims of cultivation since the long time of phytogenesis and ontogeny.

Ginkgo is the one genera with one specie and just a small one in plant taxonomy. During the long time of cultivation, by the natural hybridization and artificial selection, ginkgo has undergone some changes in fruits and leaves, and spatial isolation. It is necessary to standard classification of Ginkgo vultivars.

1. The Classification of Ginkgo Cultivars

In 1771, accoording to the specimens and the description of morphological characteristics by German doctor Kaempfer, Linnaeus named ginkgo the scientific name *Ginkgo biloba* Linn.. Scientists in the world have studied the taxonomy of ginkgo for about 100 years since then. These kinds of researches were based on different observation standards, different classification methods, different classification grades and different naming methods.

Two new taxonomic units as *Salisburia adiantifolia laciniata* Carr. and *Salisburia adiantifolia variegata* Carr. based on morphology of leaves were published by E. A. Caarriere in 1854. He determined variant of ginkgo and

纪里，中外科学家们在银杏形态分类上出现了"百家争鸣"的局面。

1862 年，Van Geert 根据枝条的自然状态发表了一个新的分类单位 Salisburia adiantifolia pendula Van Geert.（垂枝银杏）。此后，Caarriere（1867）、Beissner（1887）、Mouillefert（1898）、Henry（1906）、Makino、Rehder（1949）、Harrison（1966）等又分别对银杏品种的分类进行了相应的调整。

国内学者开展银杏品种分类的研究始于 1935 年。银杏为雌雄异株，长期以来我国以生产种子为主要栽培目的。由于种子的遗传性状相对比较稳定，因此，我国主要用银杏种核性状作为银杏品种分类的依据。曾勉在对浙江诸暨银杏调查研究基础上，根据银杏种核大小等形态指标，首次将核用品种划分为佛手银杏、梅核银杏、马铃银杏，在每个变种下再分为若干式（品种），并编制了包括 10 个品种在内的检索表。1957 年，何凤仁、赵有为将江苏泰兴的银杏鉴定为 Ginkgo biloba var. stenocarpa Hu，其中包括佛指、龙眼、野佛指、蝙蝠子、七星果等 5 个品种，编制了这 5 个品种的检索表。1978 年，郑万钧、傅立国对我国所产的 12 个主要栽培品种分别作了简要的描述记载，并按《国际栽培植物命名法规》予以命名。1979～1984 年，俞德俊、吴耕民等先后按曾勉之系统描述记载了我国各地常见的品种，将之分为佛手银杏类、梅核银杏类和马铃银杏类三大类。1989 年，何凤仁、陈鹏等在曾勉分类的基础上，提出"综合分类法"，依据核形系数（种核长宽比），将银杏栽培品种分为五大类，分别为长子（1.75～2.15）、佛手（1.45～1.75）、马铃（1.20～1.45）、梅核（1.20～1.45）和圆子（0.90～1.20）。

---

established foundation of ginkgo taxonomy. For almost a century, scientists in China and other countries in the world have carried various researches on the morphological classification of ginkgo.

A new taxonomic unit as *Salisburia adiantifolia pendula* Van Geert based on the natural status of branch was published by Van Geert in 1862. Since then, ginkgo cultivars were enriched and homologous rectified by Caarricre (1867), Beissncr (1887), Mouillefert (1898), Henry (1906), Makino, Rehder (1949) and Harrison (1966).

The study of taxonomy of ginkgo cultivar dates from 1935 in China. For a long time, farmers plant ginkgo trees for nut-producing as main cultivation purpose. Nut characters were used as main classification basis for ginkgo, because genetic characters of ginkgo fruit are comparatively stable. Three new variations (Foshou, Meihe, Maling) and a key to ten Ginkgo cultivars were published by Zeng Mian based on morphological characters of ginkgo nuts in Zhuji, Zhejiang Province. A key to five cultivars(Fozhi, Longyan, Yefozhi, Bianfuzi, and Qixingguo)in Taixing, Jiangsu Province were published by He Fengren and Zhao Youwei in 1957, and identified Ginkgo in Taixing as *Ginkgo biloba* var. *stenocarpa*. Twelve cultivars were described and named based on the ICNCP by Zheng Wanjun and Fu Liguo in 1978. The common ginkgo cultivars were described, recorded and classified as Foshou group, Meihe group and Maling group by Yu Dejun and Wu Gengmin from 1979 to 1984. A comprehensive classification method was proposed by He Fengren and Chen Peng based on the Zenmian's research in 1989, they classified

第 2 章 银杏品种分类依据
Chapter 2　Classification Basis of *Ginkgo Cultivars*

何凤仁提出的综合分类法是目前我国较为流行的分类系统，但对于以核形系数为一级分类单元分成的五大类中，马铃和梅核这两大类群的种核核形系数长宽比都是 1.45～1.45，这样在科学研究与生产实践中存在诸多不便。另外，曹福亮、桂仁意等对 44 个银杏品种种子性状的 11 个性状指标进行研究，综合产生分别代表银杏种子的种（核）宽和重、种（核）长及出核率等信息的 3 个主成分，进行聚类分析，以距离值 0.85 为标准，可以将银杏主要栽培品种划分为四大类群。因此，有必要把这两大类群合并成一大类群称为中子类，便于开展研究，以及品种的推广和应用。南京林业大学和山东农业大学等单位在 20 世纪 80 年代末率先开展核用、叶用、花粉用、观赏用等品种的分类和选育，取得了一批创新成果。

传统的植物分类方法是建立在对个体性状描述的宏观观测水平上，往往受环境因素、发育阶段等条件的限制，而利用分子标记手段和传统的形态学手段相结合的方式，有望弥补单纯依靠形态进行分类的缺陷。DNA 分子标记技术应用于银杏分类研究的时间并不长，但其在这一领域的成果却是有目共睹的。曹福亮（2005）以 44 个银杏主要栽培品种为材料，应用 ISSR 分子标记为手段，利用 5 个标记所产生的 16 个多肽位点绘制了 44 个银杏栽培品种的 ISSR 指纹图谱，估算出的平均有效等位基因数目为 1.730 7，基因多样度为 0.410 1，Shannon 信息指数为 0.596 3（图 2.1）。

银杏雄株的分类，目前还是一个空白。银杏的雄株在冠形、枝条、叶片、花期

the cultivars into five groups as follows: Changzi (1.75–2.15), Foshou (1.45–1.75), Maling (1.20–1.45) and Meihe (1.20–1.45), Yuanzi (0.90–1.20).

At present, the comprehensive classification method supported by He Fengren is comparative epidemic classification in China, based on the nut morphology coefficient, it classified the cultivars into five groups, but Maling group and Meihe group (1.20–1.45) are difficult to differentiate the two groups in study and production practice. In order to popularize and apply conveniently of cultivars of ginkgo, in our study, we suggest to combine the two groups as one

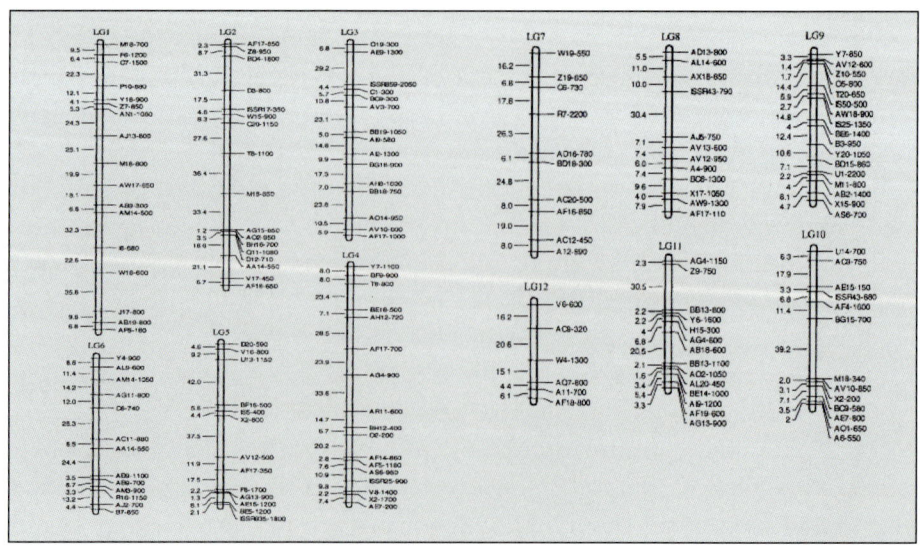

图 2.1
银杏遗传连锁图谱
Fig. 2.1
Genetic Linkage Map of Ginkgo

等方面有许多不同的类型,雄株品种的选育将有助于开展有性杂交和人工辅助授粉及扩大雄株的应用范围。美国曾有把银杏的雄株分为'金秋'('Autumn Gold')、'湖景'('Lakeview')、'五月田野'('Mayfield')、'圣诞老人'('Santa Claus')、'卫兵'('Sentry')等5个品种的报道,但未说明品种分类的依据、标准和方法。曹福亮(2006)利用 RAPD 分子标记技术对来自11个省份的银杏雄株种质资源的遗传多样性进行了研究发现,银杏雄株具有较丰富的遗传多样性,其平均有效等位基因数为1.644 0,平均基因多样度为0.375 2,平均 Shannon 信息指数为0.556 2。

遗传连锁图谱构建是基因组研究中的重要环节,是基因定位与克隆乃至基因组结构与功能研究的基础。曹福亮和桂仁意(2004)首次构建了较高饱和度的银杏遗传连锁图谱,在181个符合1∶1分离比的位点中,共有164个位点被分配到12个连锁群中,整个连锁图谱的遗传跨度为1 742.20 cm,覆盖整个基因组大小的79.2%。连锁图谱标记间的平均距离为261.2 cm,最小的连锁群图距为62.4 cm。其中,相邻连锁位点间图距最大的为42 cm,大于40 cm 的作图盲区也只一个。

2. 银杏品种鉴别

继1771年植物学家林奈(Linnaeus)正式命名银杏的学名为 *Ginkgo biloba* Linn. 之后,银杏引起了各国植物学家的关注,许多外国学者陆续到中国进行考察。但由于

---

group - Zhongzi group.

Cao Fuliang and Gui Renyi studied 11 indexes of ginkgo nuts from 44 main ginkgo cultivars. Three principal components were determined, including the width, weight and the length of fruit and nut, percent of Nut to Fruit(PNF). The main ginkgo cultivars were classified into four groups by the cluster analysis of the 3 principal components based on the distance value 0.85. They also concluded that it is conducive to selection and breeding ginkgo cultivars by using the characteristic indexes such as the stalk length, fruit weight, nut length, nut width, etc.

No work had been done on the taxonomy of male ginkgo. The male trees have different types in crown form, branches, leaves and florescence. The selection and breeding of male will be helpful to sexual hybridization, artificial pollination and enlarging application range. American researchers divided male ginkgo into five cultivars as 'Autumn Gold', 'Lakeview', 'Mayfield', 'Santa Claus' and 'Sentry', but did not publish the basis, standards and methods.

2. The Identification of Ginkgo Cultivars

Ginkgo biloba L, was given by the famous Swedish botanist Linnaeus in 1771. Since then, ginkgo evoked some botanists' attention and some of them came to China to study ginkgo. Ginkgo is a kind of giant woody plant, there were no management on ginkgo and it lived in half-wild state for a long time, so most of the botanists haven't confirmed ginkgo as a kind of cultivated plant. In history literature, ginkgo was recorded as one specie and no secondary taxonomic grade of ginkgo. When botanists came to China, they found some variant types

第 2 章 银杏品种分类依据
Chapter 2 Classification Basis of *Ginkgo Cultivars*

银杏是一种高大的木本植物，在以往的时间里多因管理粗放呈现半野生状态，所以长期以来植物界未将其作为栽培植物看待，且在历史文献中均将银杏记为一种，种以下未再设立分类等级。当许多国外植物学家到中国了解银杏生长状况和采集银杏标本的同时，发现了银杏的一些变异类型，如裂叶银杏、斑叶银杏、黄叶银杏、垂枝银杏等。对这些变异类型从植物分类的角度出发将其列为变种或变型。直到 1966 年，英国植物学家赫尔逊（S. G. Harrison）才根据《国际栽培植物命名法规》对品种（栽培变种）所下的定义将银杏的这些变异类型全部划归为品种等级。1978 年，《中国植物志·第七卷》，对银杏种级之下的所有变异类型也均列入品种等级。

1935 年，中国的果树学家曾勉在调查浙江诸暨的银杏时，根据银杏的种实和种核也曾将银杏划为 3 个变种，变种之下又分列了 10 个品种。这一分类方式就目前来看虽然不尽符合植物命名法规或栽培植物命名法规的要求，但他的功绩却是不可埋没的，一是他首先提出了银杏的品种问题；二是这一分类方法只要稍加更动（将"变种"一词改为"类别"），在应用上仍有许多方便之处。

银杏品种数量繁多，《中国果树志·银杏卷》（1993）记载的正式定名的中国银杏品种已达 46 个，且每年都有新的品种不断涌现。对这些品种的鉴别依据于银杏植株的表型特征。如核用品种以种子（种实或种核）的性状指标（大小、形状等），观赏品种以枝、叶的性状指标来鉴定等。然而大多数表型性状容易受环境条件的影响，因而地区间、年度间表现一定程度的变化，从而影响了品种遗传性描述和品种间遗传变异性评价的准确性。因此，对品种的遗传本质（包括典型性、稳定性和特异性）的客观描述就显得十分必要。

---

of ginkgo, such as *Salisburia adiantifolia laciniata* Carr., *Salisburia adiantifolia variegata* Carr., *Salisburia adiantifolia pendula* and these variant types were classified as varieties or variants based on plant taxonomy. These variant types were classified as cultivar by Botanist S. G. Harrison based on the definitions of cultivar in the ICNCP in 1966. All the variant types of ginkgo were classified as cultivars in the *Seventh Volume of Flora of China* in 1978.

Three new varieties (Foshou, Meihe, Maling) and ten Ginkgo cultivars were published by Zeng Mian based on characters of ginkgo fruit and nut in Zhuji, Zhejiang Province in 1935. Zengmian's taxonomic methods did not conform to the ICNCP or cultivated plants naming laws totally, but he made great contribution to ginkgo taxonomy. Firstly, he proposed question of the cultivar of ginkgo; secondly, the taxonomic methods was convenient, just need to change the word varieties to category.

There are so many ginkgo cultivars in China and forty six Chinese ginkgo cultivars were introduced in the *Ginkgo Volume of Fruit Flora Sinensis*. The classification basis of these cultivars is based on the morphological characteristics. For examples, the ginkgo cultivars for nut production purpose are defined by the morphological characteristics of fruits or nuts, the ginkgo cultivars for leaf-producing purpose are defined by the morphological of branches and leaves. Most of the morphological characteristics are easy to be changed by the environment, there are some changes in different places and different times, the changes

纵观银杏品种的由来与发展、品种演化过程和演进途径，结合前人已经进行的工作，银杏的品种分类体系应完全按照《国际植物命名法规》ICNCP 的规定，在明确品种所属的植物分类等级的"种系"的前提下，采用品种群、品种（含嫁接嵌合体）两级分类单位。由于银杏是单属种植物，银杏品种分类最好包括：品种群、品种。

ICNCP 第 3.1 条中明确规定：品种群是基于一定相似性的品种、植物个体或植物集合体的正式类级。组成并维持一个品种群的标准，因不同使用者的目的而异。

品种是栽培植物品种分类的最基本单位。按照 ICNCP 的规定，栽培品种是指具有一致而稳定的明显区别特征，而且采用适当的方式繁殖（有性或无性）后，这些区别特征仍能保持下来的一个栽培植物分类单位，不管这些区别特征是形态、生理、细胞、生化方面的还是其他方面的。

结合南京林业大学银杏课题组的最新研究成果，同时参照前人对银杏品种分类的方法以及考虑到科研生产的需要，我们提出了基于"功能类－品种群－品种"为核心内容的三级分类体系。该分类体系是根据培养目的将银杏分为核用、叶用、花粉用、观赏用、材用等五大类，在此基础上对于核用品种来说，根据种核形态重要指标，将核用银杏品种分为长子品种群、佛指品种群、中子品种群和圆子品种群等四个品种群，最后把每一个品种群又区分出不同的品种。而对于观赏用、花粉用、叶用、材用等四大功能类的银杏来说，目前没有区分到"品种群"一级，而是直接根据一些重要形态、生理、生化等指标，区分出品种或无性系。

influence accuracy of description of genetic in cultivar and description of genetic variability between different cultivars. So, it is necessary to describe the genetic nature such as typicality, stability and specificity.

Taking a panoramic view of the origin and development, the evolution process, the evolution means of *Ginkgo biloba* cultivars, and combining with predecessors's work, we think that *Ginkgo biloba* cultivar classification system should conform to the provisions of the ICNCP, and adopt two grades, cultivar group and cultivar.

Regulations were clearly stipulated in Article 3.1 of ICNCP that the group is an official grade based on a certain similarity of cultivars and plant individuals or plant categories. The standard confirming a cultivar group is different from the purposes of different users.

Cultivar is the basic classification unit of cultivated plants. According to ICNCP, the cultivar is a unit with coherent and stable characteristics of clear distinction, and these distinctive features can maintain a separation after the use of appropriate methods of reproduction (sexual and asexual), no matter whether it has the distinctive features in morphology, physiology, cells, or other aspects of the biochemistry.

Based on the purposes of cultivation, Ginkgo tree species was divided into 5 Sections (Nut-producing, Ornamental Purpose, Leaf-producing, Pollen-producing and Timber-producing). For Nut-producing Section, it was further divided into 4 groups (Changzi group, Fozhi group, Zhongzi group and Yuanzi group). For each group, there are many cultivars.

广西桂林海洋乡梦幻秋色
Dreamy Autumn Tints in Haiyang, Guilin, Guangxi

第 3 章
Chapter 3

银杏品种检索表及品种图谱

Key to Cultivars of Ginkgo biloba L. and Illustration of Ginkgo Cultivars

3.1 银杏核用品种及无性系

对于以生产种实为目的的大部分雌株，考虑到种实、种核等与繁殖器官有关特征的稳定性和保守性，银杏核用品种群检索表的编制重点依据种核的长宽比（核形系数 RLW：ratio of nut length to width）将银杏以核用为目的的品种划分为 4 个品种群，即长子品种群（RLW 大于 1.75）、佛指品种群（RLW 介于 1.50～1.75）、中子品种群（RLW 介于 1.30～1.50）、圆子品种群（RLW 小于 1.30）（见银杏核用品种群检索表）。每个品种群均具有明显特征，而品种群内各品种既有其特异性，又具有共同特征。另外品种群类的品种区分主要依据种核形状、种实特征、种柄特征等指标来进行。4 个品种群的划分与传统的分类习惯是一致的。

银杏核用品种群检索表

1. 种核长与宽的比值大于 1.75（±0.05），种核纵横轴的交叉点位于纵线的中心··1. 长子品种群
1. 种核长与宽的比值小于 1.75（±0.05）
  2. 种核长与宽的比值介于 1.50（±0.05）与 1.75（±0.05）之间，种核纵横轴的交叉点位于纵线 1/3 处···2. 佛指品种群
  2. 种核长与宽的比值小于 1.50（±0.05）
    3. 种核长与宽的比值介于 1.30（±0.05）与 1.50（±0.05）之间···················3. 中子品种群
      4. 种核纵横轴线交叉点位于纵线的 2/5 处 ································（马铃亚品种群）
      4. 种核纵横轴线交叉点位于纵线的 1/2 处 ································（梅核亚品种群）
    3. 种核长与宽的比值小于 1.30（±0.05），种核纵横轴的交叉点位于纵线的中心··4. 圆子品种群

3.1 Ginkgo Cultivars and Clones for Nut-producing

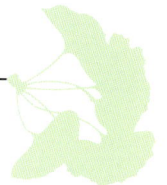

For nut-producing cultivars, the ratio of nut length to width, nut type, fruit type, stalk and other reproductive characteristics are more conservative than those of vegetative organs. The ginkgo cultivars for nut production purpose can be divided into 4 groups, based on the ratio of nut length to its width (RLW), the figure of nut, the character of fruit and the stalk. They are Changzi Group(with the ratio >1.75), Fozhi Group(with the ratio 1.50–1.75), Zhongzi Group(with the ratio 1.30–1.50) and Yuanzi Group(with the ratio <1.30), respectively. Each group has its own clear distinction. Cultivars are different from each other in each group. Key to groups of Ginkgo cultivars for Nut-producing is as follows.

Key to Groups of Ginkgo Cultivars for Nut-producing

1. Ratio of length to width beyond 1.75 (±0.05), the intersection of ordinate axis and horizontal axis of nut site one half of ordinate axis .... 1. Changzi Group
1. Ratio of length to width less than 1.75 (±0.05)
   2. Ratio of length to width, range from 1.50 (±0.05) to 1.75 (±0.05), the intersection of ordinate axis and horizontal axis of nut site one third of ordinate axis ... 2. Fozhi Group
   2. Ratio of length to width less than 1.50 (±0.05)
      3. Ratio of length to width, range from 1.30(±0.05) to 1.50 (±0.05) .... 3. Zhongzi Group
         4. The intersection of ordinate axis and horizontal axis of nut site two fifth of ordinate axis .... (Maling Sub-group)
         4. The intersection of ordinate axis and horizontal axis of nut site one half of ordinate axis ....(Meihe Sub-group)
      3. Ratio of length to width less than 1.30 (±0.05), the intersection of ordinate axis and horizontal axis of nut site one half of ordinate axis .... 4. Yuanzi Group

3.1.1　长子品种群

将银杏种核长与宽的比值大于 1.75（±0.05），种核纵横轴的交叉点位于纵线的中心的银杏品种划分为长子品种群，共有 5 个优良品种及 15 个优良无性系。本节重点介绍这 5 个优良品种特性，另有 15 个优良无性系只以表格和照片的形式做简要介绍。

3.1.1.1　长子品种群分类检索表

长子品种群检索表

1. 种实熟时表面无光泽，被白粉；种核先端光滑，无明显条纹
　2. 种实先端秃圆，顶凹陷，种核有背腹之分
　　3. 种实外种皮熟时青黄色，油胞较多 …………………………………………… 1.'橄榄果'
　　3. 种实外种皮熟时橙黄色，无油胞 ………………………………………………… 2.'圆枣佛手'
　2. 种实先端秃圆，顶具小尖，种核无背腹之分
　　4. 蒂盘圆形或近椭圆形，表面凸凹不平，周缘略凹陷…………………………… 3.'天目长籽'
　　4. 蒂盘圆或椭圆，周缘不整，略高于种皮……………………………………… 4.'九甫长籽'
1. 种实熟时表面有光泽，基本无白粉；种核先端粗糙，有明显条纹 ……………………… 5.'大金坠'

3.1.1 Changzi Group

There are 5 cultivars and 15 clones of Changzi Group, according to the ratio of length to width beyond 1.75 (±0.05) and the intersection of ordinate axis and horizontal axis of nut site one half of ordinate axis.

3.1.1.1 Key to Changzi Group

Key to Changzi Group of *Ginkgo biloba* L.

1. Fruits lusterless when ripe, with glaicous; nuts glazed and without conspicuous stria at the apex
   2. Fruits blunt and round at the tip, emarginated at the apex, nuts asymmetrical
      3. Fruits with many oil sac and green yellow when ripe..........................1. 'Ganlan Guo'
      3. Fruits without oil sac and orange yellow when ripe.....................2. 'Yuanzao Foshou'
   2. Fruits blunt and round at the tip, with cusps at the apex, nuts symmetrical
      4. The base of stalk round or approximate elliptic, surface scraggly, and circumferentia slight concave..3. 'Tianmu Changzi'
      4. The base of stalk round or elliptic, circumferentia slight protuberant...4. 'Jiufu Changzi'
1. Fruits lustrous when ripe, without glaicous; nuts rough and with conspicuous stria at the apex ....5. 'Da Jinzhui'

广西桂林海洋乡银杏林
Ginkgo Trees in Haiyang, Guilin, Guangxi

第3章 银杏品种检索表及品种图谱
Chapter 3 Key to Cultivars of *Ginkgo biloba* L. and Illustration of Ginkgo Cultivars

3.1.1.2　长子品种群优良品种
分别对长子品种群5个优良品种进行详细介绍。

3.1.1.2　Cultivars of Changzi Group
Illustrat 5 cultivars of Changzi Group in detail.

1　'橄榄果'

又称'橄榄佛手'、'大钻头'。本品种的种实与种核均似橄榄，故名。核形系数为1.80。150年生大树，树高17.5 m，胸径61.6 cm，冠幅8.0 m×9.1 m，树冠圆锥形。种实长卵圆形，种实先端微圆秃，顶凹陷，具小尖，珠孔迹明显。中部以下稍狭，蒂盘多椭圆形，稍偏斜，周缘不整，微有凹陷。果柄长约3.6～4.4 cm，略弯曲，细弱，斜生。熟时青黄色，油胞较多，白粉亦多。种核长卵圆形或长椭圆形，先端突尖，珠孔迹明显。下部稍狭，基部具尖，两维管束迹迹点小，相距可达3.2 mm。两侧棱明显，较宽，上部宽扁，呈翼状，中下部逐渐消逝。有背腹之分。种核大小为2.7 cm×1.5 cm×1.3 cm，千粒重2 400 g，出核率（PNF：percent of nut to fruit）24.7%，出仁率（PKN：percent of kernel to nut）77.4%。目前主要分布于广西桂林的灵川和兴安。

'Ganlan Guo'

'Ganlan Foshou', 'Da Zuantou'.

Fruits and nuts olive-shaped. RLW 1.80. The tree 150 years old, up to 17.5m high, 61.6cm dbh. Crown conoid, 8.0 m×9.1 m. Fruits long ovoid, broad round and emarginated with cusps and conspicuous micropylar at the apex. The base of stalk ellipse, slight oblique, surface scraggly, and the circumferentia slight concave; stalk slight oblique, 3.6–4.4 cm long. Fruits green yellow, with many oil sac and glaicous when ripe. Nuts asymmetrical, long ovoid or long ellipsoid, with conspicuous micropylar at the apex, with cusps and two little vascular bundle scars at the base. Raphe conspicuous from the apex to the midst, wing-shaped. Size of nut is 2.7 cm×1.5 cm×1.3 cm, 1 000-grain weight of 2 400 g, PNF 24.7% and PKN 77.4% normally. Mainly cultivated in Guilin, Guangxi.

图3.1　'橄榄果'
①广西桂林海洋乡50年生实生大树；
②结果枝生长状态

Fig. 3.1　'Ganlan Guo' (Mother Plant)
① 50 years old tree from seedling in Guilin, Guangxi,
② Bearing shoots

银杏核用品种——长子品种群
Changzi Group for Nut-producing

图 3.2 '橄榄果'
① 种实和种核；
② 种实和叶子；
③ 结果枝蜡叶标本

Fig. 3.2 'Ganlan Guo'
① Fruits and nuts,
② Fruits and leaves,
③ Wax-leaf specimens of bearing shoots

2 '圆枣佛手'

又称'枣子佛手'、'枣子果'。本品种之种实果形近似大枣，故名。核形系数1.93。树冠塔形。种实长椭圆形，先端圆秃，珠孔迹横宽呈"一"字形凹入。基部稍狭，蒂盘圆形或椭圆形，稍偏斜或不偏斜，周缘不整，略凹陷。果柄略偏斜，粗而长，长3.8~4.2 cm。熟时深橙黄色，满被薄白粉。种核长倒卵圆形，两端均尖。先端钝尖，珠孔迹明显。基部稍狭，尾端秃尖，两维管束迹迹点极小，相距较近，仅1.0~2.0 mm，偶见合为一体而呈宽扁突起状。两侧棱明显，上半部尤显，中下部逐渐消逝。有背腹之分，但不太明显。千粒重2 070 g。种核大小为2.7 cm×1.4 cm×1.2 cm，出核率24.7%，出仁率71.0%~75.3%。位于广西桂林灵川的海洋、潮田，兴安的漠川、白石，全州的焦江、安和等。

'Yuanzao Foshou'

'Zaozi Foshou', 'Zaozi Guo'.

Fruits jujube-shaped. RLW 1.93. Crown tower-shaped. Fruits long ellipsoid, broad round and emarginated, with conspicuous micropylar at the apex. The base of stalk round or ellipse, slight oblique, surface scraggly, and the circumferentia slight concave; stalk slight oblique, 3.8–4.2 cm long. Fruits orange yellow, covered with glaicous when ripe. Nuts asymmetrical, obovoid with conspicuous micropylar at the apex, with cusps and two little vascular bundle scars at the base with distance 1.0–2.0 mm or adnate. Raphe conspicuous from the apex to the midst, wing-shaped. Size of nut is 2.7 cm×1.4 cm×1.2 cm. 1 000-grain weight of 2 070 g. PNF 24.7% and PKN 71.0%–75.3% normally. Mainly cultivated in Haiyang, Chaotian, Mochuan, Baishi of Guilin, Guangxi.

图 3.3
'圆枣佛手'
① 广西桂林海洋乡40年生实生大树；
② 结果枝生长状态

Fig. 3.3 'Yuanzao Foshou' (Mother Plant)
① 40 years old tree from seedling in Guilin, Guangxi，
② Bearing shoots

银杏核用品种——长子品种群
Changzi Group for Nut-producing

图3.4 '圆枣佛手'
① 种实和种核；
② 种实和叶子；
③ 结果枝蜡叶标本

Fig. 3.4 'Yuanzao Foshou'
① Fruits and nuts,
② Fruits and leaves,
③ Wax-leaf specimens of bearing shoots

3 '天目长籽'

核形系数1.88。浙江天目山禅源寺内约200年生之银杏，其树高为15 m，干高4 m，胸径88.2 cm，冠幅10 m×10 m。树冠圆头形，主干明显，生长势不强，发枝力较弱。叶多三角形、扇形和截形，具明显中裂，但裂口较浅，深约2.0~6.0 mm。扇形叶长6.0 cm，宽达7.8 cm，叶片大而厚。叶柄长平均4.50 (2.0~7.5) cm。种实长卵圆形，先端秃圆，具小尖，基部蒂盘圆形或近椭圆形，表面凹凸不平，周缘略凹陷。果柄长约2.7 cm。熟时黄色，被薄白粉。多双果。种核长卵圆形。先端圆钝，具秃尖，珠孔迹明显，基部两束迹迹点明显，相距较近，约1.2~2 mm。两侧棱线明显，但不成翼状，仅近基部处隐约可见。种核大小为3.00 cm×1.60 cm×1.57 cm，千粒重2 390~2 630 g，出核率24.0%，出仁率74.0%。

'Tianmu Changzi'

RLW 1.88.The tree grows in the Chanyuan Temple, Tianmu Mountain, Zhejiang. 200 years old, up to 15 m high, 0.88 m dbh; crown spheroidal, 10m×10m.Leaves fan-shaped, triangle, truncate, 2-lobed, 2.0–6.0 mm. fan-shaped leaf, 6.0cm long, 7.8 cm wide. Leaves with petiole 4.50 (2.0–7.5) cm. Fruits long ovoid, broad round and emarginated, with cusps at the apex. The base of stalk round or ellipse, surface scraggly, and the circumferentia slight concave, stalk slight oblique, 2.7 cm long. Fruits yellow, covered thin glaicous when ripe. Nuts asymmetrical, long ovoid, broad round and emarginated, with cusps and conspicuous micropylar at the apex, with two conspicuous little vascular bundle scars at the base with distance 1.2–2 mm. Raphe conspicuous in the base, not wing-shaped. Size of nut is 3.00 cm×1.60 cm×1.57 cm. 1 000-grain weight of 2 390–2 630 g, PNF 24% and PKN 74% normally.

图 3.5 '天目长籽'
①浙江天目山禅源寺前 200 年生实生大树；
②结果枝生长状态

Fig. 3.5 'Tianmu Changzi' (Mother Plant)
① 200-year-old tree from seedling in Tianmu Mountain, Zhejiang,
② Bearing shoots

银杏核用品种——长子品种群
Changzi Group for Nut-producing

图 3.6 '天目长籽'
① 种实和种核；
② 种实和叶子；
③ 结果枝蜡叶标本

Fig. 3.6 'Tianmu Changzi'
① Fruits and nuts,
② Fruits and leaves,
③ Wax-leaf specimens of bearing shoots

4 '九甫长籽'

核形系数1.89。母树树高15.0 m，冠幅14.0 m×14.0 m，树冠半圆头形。种实长卵圆形，先端浑圆，具小尖。基部圆筒状，蒂盘圆或椭圆，周缘不整，略高于种皮。果柄长4.2 cm。熟时黄色，被薄白粉。种核长椭圆形，两端基本相等，先端稍圆秃，具小尖，基部束迹迹点小，相距近1.8～2.0 mm。两侧棱自上至下均不明显，但近基部处仅隐约可见。种核大小为2.84 cm×1.50 cm×1.28 cm，千粒重2 270～2 630 g，每千克粒数380～441粒。出核率23.5%，出仁率72.1%。主要产于浙江临安昌化。

'Jiufu Changzi'

RLW 1.89. The tree up to 15.0m high, 14.0 m×14.0 m crown. Crown spheroidal. Fruits long ovoid, broad round, with cusps at the apex. The base of stalk round or ellipse, surface scraggly, and the circumferentia slight protuberant; stalk 4.2 cm long. Fruits yellow, covered thin glaicous when ripe. Nuts asymmetrical, long ovoid, broad round, with cusps at the apex, with two conspicuous little vascular bundle scars at the base with distance 1.8–2.0 mm. Raphe conspicuous in the base. Size of nut is 2.84 cm×1.50 cm×1.28 cm. 1 000-grain weight of 2 270–2 630 g. PNF 23.5% and PKN 72.1% normally. Mainly cultivated in Changhua, Lin'an, Zhejiang.

图 3.7 '九甫长籽'
结果枝生长状态

Fig. 3.7 'Jiufu Changzi'
Bearing shoots

银杏核用品种——长子品种群
Changzi Group for Nut-producing

图3.8 '九甫长籽'
① 种实和种核；
② 种实和叶子；
③ 结果枝蜡叶标本

Fig. 3.8 'Jiufu Changzi'
① Fruits and nuts,
② Fruits and leaves,
③ Wax-leaf specimens of bearing shoots

5 '大金坠'

又称'长白果'。本品种之特点为种实与种核均似妇女之耳坠,故名。核形系数为1.84。树冠多圆锥形或扁球形。种实倒卵圆形或长椭圆形,种实先端突出,顶具尖或不具尖。基部略偏斜,蒂盘圆形,较厚,凸出于种皮之上或与种皮相平。果柄长3.0～3.5 cm左右,柄粗壮、直立、不弯曲。种实最宽处位于中部以下。熟时金黄色有光泽,基本无白粉。种核长卵形,先端粗糙,有明显的条纹,具钝尖。基部尖或略钝圆。两端窄狭,最宽处位于中部以下。种核大小为2.40 cm×1.30 cm×1.80 cm,千粒重1 910～3 440 kg。出核率26%左右,出仁率约80%。主要分布于山东郯城的港上、新村、胜利等沿沂武河两岸的乡镇。

'Da Jinzhui'

'Chang Baiguo'.
Fruits and nuts similar to the shape of earbob. RLW 1.84. Crown conoid or oblate spheroid. Fruits obovoid or ellipsoid, with cusps or without cusp at the apex. The base of stalk round, thicker, slight protuberant; stalk erect 3–3.5 cm long; golden brown and lustrous when ripe, without glaicous. Nuts long ovoid, rough and with conspicuous stria at the apex; blunt and round at the base, the apex and the base narrow, the most width of nut beneath midst. Size of nut is 2.40 cm×1.30 cm×1.80 cm. 1 000-grain weight of 2 675 (1 910–3 440) g. PNF 26% and PKN 80% normally. Mainly cultivated in Gangshang, Xincun, Shengli of Tancheng, Shandong.

图 3.9 '大金坠'
山东郯城清泉寺林场银杏种质基因库

Fig. 3.9 'Da Jinzhui' Bearing shoots in Ginkgo resources gene bank in Tancheng, Shandong

银杏核用品种——长子品种群
Changzi Group for Nut-producing

图 3.10 '大金坠'
① 种实和种核；
② 种实和叶子；
③ 结果枝蜡叶标本

Fig. 3.10 'Da Jinzhui'
① Fruits and nuts,
② Fruits and leaves,
③ Wax-leaf specimens of bearing shoots

3.1.1.3 长子品种群优良无性系

本节分别对长子品种群中 15 个优良无性系进行简要介绍，见表 3.1。

3.1.1.3 Clones of Changzi Group

Table 3.1 illustrated 15 clones of Changzi group in brief.

表 3.1 长子品种群优良无性系
Table 3.1 A list of Some Ginkgo Clones of Changzi Group

编号 Number	名称 Name	核形系数 RLW	产地 Distribution
1	'长兴 1 号' 'Changxing-1'	1.76	浙江长兴 Changxing, Zhejiang
2	'泰兴 4 号' 'Taixing-4'	1.77	扬州大学 Yangzhou University
3	'长兴 3 号' 'Changxing-3'	1.78	浙江长兴 Changxing, Zhejiang
4	'长兴 4 号' 'Changxing-4'	1.78	浙江长兴 Changxing, Zhejiang
5	'安陆 A11' 'Anlu-A11'	1.79	湖北安陆 Anlu, Hubei
6	'新村 231 号' 'Xincun-231'	1.82	山东郯城新村 Xincun, Tancheng, Shandong
7	'胜利 102 号' 'Shengli-102'	1.83	山东郯城胜利 Shengli, Tancheng, Shandong
8	'灵川 F9' 'Lingchuan-F9'	1.85	广西灵川 Lingchuan, Guangxi
9	'长兴 F13' 'Changxing-F13'	1.85	浙江长兴 Changxing, Zhejiang
10	'新村 402 号' 'Xincun-402'	1.86	山东郯城新村 Xincun, Tancheng, Shandong
11	'新村 222 号' 'Xincun-222'	1.86	山东郯城新村 Xincun, Tancheng, Shandong
12	'安吉 F4' 'Anji-F4'	1.86	浙江安吉 Anji, Zhejiang
13	'东山 F15' 'Dongshan-F15'	1.91	江苏吴县东山 Dongshan, Wuxian, Jiangsu
14	'重坊 176 号' 'Chongfang-176'	2.00	山东郯城重坊 Chongfang, Tancheng, Shandong
15	'新村 203 号' 'Xincun-203'	2.21	山东郯城新村 Xincun, Tancheng, Shandong

银杏核用品种——长子品种群
Changzi Group for Nut-producing

'长兴1号' 'Changxing-1'

图 3.11
'长兴1号'
结果枝生长状态

Fig. 3.11
'**Changxing-1**'
Bearing shoots

图 3.12
'长兴1号'
①种实和种核；
②种实和叶子；
③结果枝蜡叶标本

Fig. 3.12
'**Changxing-1**'
① Fruits and nuts,
② Fruits and leaves,
③ Wax-leaf specimens of bearing shoots

35

第3章 银杏品种检索表及品种图谱
Chapter 3 Key to Cultivars of *Ginkgo biloba* L. and Illustration of Ginkgo Cultivars

'泰兴 4 号' 'Taixing-4'

图 3.13
'泰兴 4 号'
结果枝生长状态

Fig. 3.13
'Taixing-4'
Bearing shoots

图 3.14
'泰兴 4 号'
① 结果枝蜡叶标本；
② 种实和种核；
③ 种实和叶子

Fig. 3.14
'Taixing-4'
① Wax-leaf specimens of bearing shoots,
② Fruits and nuts,
③ Fruits and leaves

银杏核用品种——长子品种群
Changzi Group for Nut-producing

'长兴3号' 'Changxing-3'

图 3.15
'长兴3号'
结果枝生长状态

Fig. 3.15
'Changxing-3'
Bearing shoots

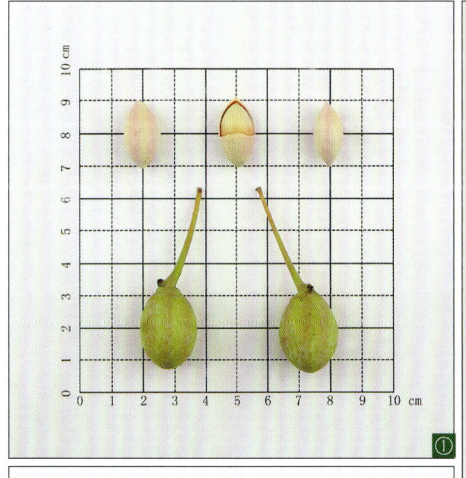

图 3.16
'长兴3号'
①种实和种核；
②种实和叶子；
③结果枝蜡叶标本

Fig. 3.16
'Changxing-3'
① Fruits and nuts,
② Fruits and leaves,
③ Wax-leaf specimens of bearing shoots

第3章 银杏品种检索表及品种图谱
Chapter 3 Key to Cultivars of *Ginkgo biloba* L. and Illustration of Ginkgo Cultivars

'长兴4号' 'Changxing-4'

图 3.17
'长兴4号'
结果枝生长状态

Fig. 3.17
'Changxing-4'
Bearing shoots

图 3.18
'长兴4号'
① 结果枝蜡叶标本；
② 种实和种核；
③ 种实和叶子

Fig. 3.18
'Changxing-4'
① Wax-leaf specimens of bearing shoots,
② Fruits and nuts,
③ Fruits and leaves

银杏核用品种——长子品种群
Changzi Group for Nut-producing

'安陆 A11' 'Anlu-A11'

图 3.19
'安陆 A11'
结果枝生长状态

Fig. 3.19
'Anlu-A11'
Bearing shoots

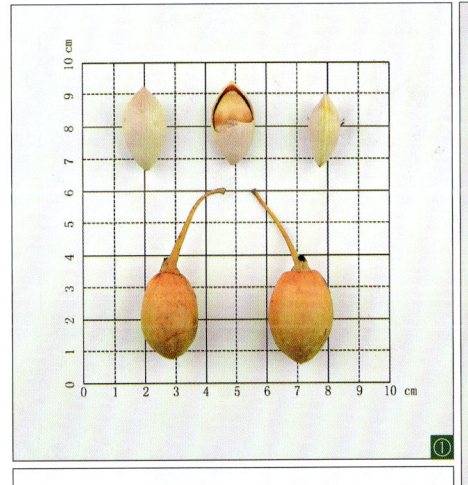

图 3.20
'安陆 A11'
①种实和种核；
②种实和叶子；
③结果枝蜡叶标本

Fig. 3.20
'Anlu-A11'
① Fruits and nuts,
② Fruits and leaves,
③ Wax-leaf specimens of bearing shoots

第3章 银杏品种检索表及品种图谱
Chapter 3　Key to Cultivars of *Ginkgo biloba* L. and Illustration of Ginkgo Cultivars

'新村 231 号' 'Xincun-231'

图 3.21
'新村 231 号'
结果枝生长状态

Fig. 3.21
'Xincun-231'
Bearing shoots

图 3.22
'新村 231 号'
①结果枝蜡叶标本；
②种实和种核；
③种实和叶子

Fig. 3.22
'Xincun-231'
① Wax-leaf specimens of bearing shoots,
② Fruits and nuts,
③ Fruits and leaves

'胜利 102 号' 'Shengli-102'

图 3.23
'胜利 102 号'
结果枝生长状态

Fig. 3.23
'Shengli-102'
Bearing shoots

图 3.24
'胜利 102 号'
① 种实和种核；
② 种实和叶子；
③ 结果枝蜡叶标本

Fig. 3.24
'Shengli-102'
① Fruits and nuts,
② Fruits and leaves,
③ Wax-leaf specimens of bearing shoots

'灵川 F9' 'Lingchuan-F9'

图 3.25
'灵川 F9'
结果枝生长状态

Fig. 3.25
'Lingchuan-F9'
Bearing shoots

图 3.26
'灵川 F9'
①结果枝蜡叶标本；
②种实和种核；
③种实和叶子

Fig. 3.26
'Lingchuan-F9'
① Wax-leaf specimens of bearing shoots,
② Fruits and nuts,
③ Fruits and leaves

'长兴F13' 'Changxing-F13'

图 3.27
'长兴F13'
结果枝生长状态

Fig. 3.27
'Changxing-F13'
Bearing shoots

图 3.28
'长兴F13'
① 种实和种核；
② 种实和叶子；
③ 结果枝蜡叶标本

Fig. 3.28
'Changxing-F13'
① Fruits and nuts,
② Fruits and leaves,
③ Wax-leaf specimens of bearing shoots

'新村 402 号' 'Xincun-402'

图 3.29
'新村 402 号'
结果枝生长状态

Fig. 3.29
'Xincun-402'
Bearing shoots

图 3.30
'新村 402 号'
①结果枝蜡叶标本；
②种实和种核；
③种实和叶子

Fig. 3.30
'Xincun-402'
① Wax-leaf specimens of bearing shoots,
② Fruits and nuts,
③ Fruits and leaves

银杏核用品种——长子品种群
Changzi Group for Nut-producing

'新村 222 号' 'Xincun-222'

图 3.31
'新村 222 号'
结果枝生长状态

Fig. 3.31
'Xincun-222'
Bearing shoots

图 3.32
'新村 222 号'
①种实和种核；
②种实和叶子；
③结果枝蜡叶标本

Fig. 3.32
'Xincun-222'
① Fruits and nuts,
② Fruits and leaves,
③ Wax-leaf specimens of bearing shoots

'安吉 F4' 'Anji-F4'

图 3.33
'安吉 F4'
结果枝生长状态

Fig. 3.33
'Anji-F4'
Bearing shoots

图 3.34
'安吉 F4'
①结果枝蜡叶标本；
②种实和种核；
③种实和叶子

Fig. 3.34
'Anji-F4'
① Wax-leaf specimens of bearing shoots,
② Fruits and nuts,
③ Fruits and leaves

银杏核用品种——长子品种群
Changzi Group for Nut-producing

'东山 F15' 'Dongshan-F15'

图 3.35
'东山 F15'
结果枝生长状态

Fig. 3.35
'Dongshan-F15'
Bearing shoots

图 3.36
'东山 F15'
① 种实和种核；
② 种实和叶子；
③ 结果枝蜡叶标本

Fig. 3.36
'Dongshan-F15'
① Fruits and nuts,
② Fruits and leaves,
③ Wax-leaf specimens of bearing shoots

第3章 银杏品种检索表及品种图谱
Chapter 3 Key to Cultivars of *Ginkgo biloba* L. and Illustration of Ginkgo Cultivars

'重坊176号' 'Chongfang-176'

图 3.37
'重坊176号'
结果枝生长状态

Fig. 3.37
'Chongfang-176'
Bearing shoots

图 3.38
'重坊176号'
①结果枝蜡叶标本；
②种实和种核；
③种实和叶子

Fig. 3.38
'Chongfang-176'
① Wax-leaf specimens of bearing shoots,
② Fruits and nuts,
③ Fruits and leaves

'新村 203 号' 'Xincun-203'

图 3.39
'新村 203 号'
结果枝生长状态

Fig. 3.39
'Xincun-203'
Bearing shoots

图 3.40
'新村 203 号'
①种实和种核；
②种实和叶子；
③结果枝蜡叶标本

Fig. 3.40
'Xincun-203'
① Fruits and nuts,
② Fruits and leaves,
③ Wax-leaf specimens of bearing shoots

3.1.2 佛指品种群

将银杏种核长与宽的比值介于1.50（±0.05）与1.75（±0.05）之间，种核纵横轴的交叉点位于纵线1/3处的银杏品种划分为佛指品种群，共有13个优良品种及25个优良无性系。本节内容重点介绍这13个优良品种特性，另有25个优良无性系只以表格和照片的形式作简要介绍。

3.1.2.1 佛指品种群分类检索表

佛指品种群检索表

1. 种核两面对称，背腹厚度相等
 2. 种核两面具凹陷形麻点或短条纹
  3. 种核两面具针孔状稀疏凹点，种实熟时橙黄色 ………………………………1.'七星果'
  3. 种核两面具不规则凹陷短条纹，种实熟时表面呈青绿色（绿黄色），具透明梭状油胞 ………………………………2.'青皮果'
 2. 种核两面光滑洁白，不具凹点或短条纹
  4. 种核先端有明显小尖
   5. 种核先端宽圆，顶部具小尖常凹入
    6. 种核先端尖扁而形同鸭尾，先端无尖棱 ………………………3.'鸭尾银杏'
    6. 种核先端顶尖凹陷，但可见珠孔孔迹 ………………………4.'长糯白果'
   5. 种核先端尖削，具秃尖，珠孔迹明显
    7. 种核基部维管束迹点明显
     8. 种核两侧有棱线无翼状，上下部均明显
      9. 种核长椭圆形
       10. 种实基部蒂盘近正圆，种柄着生处无隆起小阜 ………………5.'家佛指'
       10. 种实基部蒂盘长圆形或椭圆形，种柄着生处有隆起小阜 ……………………………………6.'洞庭皇'
      9. 种核倒卵圆形 ………………………………………………7.'郯魁'
     8. 种核两侧有棱线无翼状，只有中上部明显，中部以下不清晰
      11. 种实表面有疣状凸起，种核基部两维管束迹点小………8.'贵州长白果'
      11. 种实表面有明显缢痕，种核基部两维管束迹点大，有木质化横隔相连 ………………………………………………9.'黄皮果'
    7. 种核基部维管束迹合生 ………………………………………10.'金坠13号'
  4. 种核先端无明显小尖 ……………………………………………………11.'华口大白果'
1. 种核两面不对称，背圆腹薄
 12. 种核基部两维管束合生 ………………………………………………………12.'新宇'
 12. 种核基部两维管束明显大，有时相连成鸭尾状 ………………………………13.'扁佛指'

3.1.2 Fozhi Group

There are 13 cultivars and 25 clones of Fozhi group, according to the ratio of length to width, range from 1.50 (±0.05) to 1.75 (±0.05) and the intersection of ordinate axis and horizontal axis of nut situates in one third of ordinate axis.

3.1.2.1 Key to Fozhi Group

Key to Fozhi Group of *Ginkgo biloba* L.

1. Nuts symmetrical
    2. Nuts with foveolate pock or short stria
        3. Nuts with pinhole-shaped, sparse and foveolate pocks, fruits orange yellow when ripe .... 1. 'Qixing Guo'
        3. Nuts with irregular and foveolate, short stria, fruits with spindle-shaped and hyaloid oil sac, green-yellow when ripe .... 2. 'Qingpi Guo'
    2. Nuts white and lubricious, without foveolate pock or short stria
        4. Nuts with conspicuous cusps at the apex
            5. Nuts round and wide, with foveolate cusps at the apex
                6. Nuts acuminate, oblate and duck tail-shaped, without cusp at the apex .... 3. 'Yawei Yinxing'
                6. Nuts foveolate and with micropylar scar at the apex...... 4. 'Changnuo Baiguo'
            5. Nuts taper, with cusps and conspicuous micropylar scar at the apex
                7. Nuts with conspicuous vascular bundle scars at the base
                    8. Nuts with conspicuous but not wing-shaped raphe
                        9. Nuts long ellipsoid
                            10. The base of stalk long round, stalk without a protuberant abortive oval .... 5. 'Jia Fozhi'
                            10. The base of stalk long round or elliptic, stalk with a protuberant abortive oval .... 6. 'Dongting Huang'
                        9. Nuts obovoid .... 7. 'Tan Kui'
                    8. Raphe conspicuous from midst to apex
                        11. Fruits with nodule-shaped hump, nuts with small vascular bundle scars at the base .... 8. 'Guizhou Chang Baiguo'
                        11. Fruits with conspicuous constriction, nuts with larger and adnate vascular bundle scars at the base............... 9. 'Huangpi Guo'
                7. Nuts with adnate vascular bundle scars at the base................ 10. 'Jinzhui-13'
        4. Nuts without conspicuous cusp at the apex......................... 11. 'Huakou Da Baiguo'
1. Nuts asymmetrical
    12. Vascular bundle scars adnate at the base .... 12. 'Xin Yu'
    12. Vascular bundle scars conspicuous and duck tail-shaped occasionally......13. 'Bian Fozhi'

3.1.2.2 佛指品种群优良品种
分别对佛指品种群13个优良品种进行详细介绍。

3.1.2.2 Cultivars of Fozhi Group
Illustrated 13 cultivars of Fozhi Group in detail.

1 '七星果'

本品种种核两面均具针孔状稀疏分布之麻点，凹凸不一，类似群星而得名。核形系数1.73。叶多为扇形，少数为截形叶，叶有明显浅狭中裂，边缘具不明显的缺刻，叶柄短而细，叶色浓绿。七星果种实长椭圆形，先端钝圆，不具凸尖，较平阔。基部蒂盘长圆形，略歪斜，周缘稍有凹陷。熟时橙黄色，被薄白粉，上下均较胖。种核为长卵圆形，脊腹相等，核体较胖。基部两束迹明显分开，但多有横隔相连。两侧棱线明显，但不成翼状，自上而下明显。顶端孔迹突起成小尖。种核大小均匀，种核最大特点是在背腹面上有5~10个凹陷不一的麻点，种壳鱼肚白。种核大小为2.29 cm×1.32 cm×1.20 cm，千粒重2 730 g。出核率25%，出仁率81%。本品种产于江苏泰兴。

'Qixing Guo'

Nuts covered with pinhole-shaped, sparse and foveolate pocks similar to the shape of group stars. RLW1.73. Leaves fan shaped, rarely cuneat, conspicuous 2-lobed, margin not conspicuous incised, deep green, petioles short and slender. Fruits long ellipsoid, blunt and rounded, without cusps at the apex; the base of stalk long-rounded, slight oblique, margin emarginate at the base, orange yellow and glaicous when ripe. Nuts long ovoid, symmetrical, vascular bundle scars conspicuously divided at the base, raphe conspicuous, not wing-shaped, micropyla scar apiculate at the apex, foveolates 5–10. Size of nut is 2.29 cm×1.32 cm×1.20 cm. 1 000-grain weight of 2 730 g. PNF 25% and PKN 81%. Mainly cultivated in Taixing, Jiangsu.

图 3.41
'七星果'（原株）
①江苏泰州刁铺镇30年生实生大树；
②结果枝生长状态

Fig. 3.41
'Qixing Guo'
① 30 years old tree from seedling in Taizhou, Jiangsu, ② Bearing shoots

银杏核用品种——佛指品种群
Fozhi Group for Nut-producing

图 3.42 '七星果'
①种实和种核；
②种实和叶子；
③结果枝蜡叶标本

Fig. 3.42 'Qixing Guo'
① Fruits and nuts,
② Fruits and leaves,
③ Wax-leaf specimens of bearing shoots

2 '青皮果'

核形系数1.50。树冠圆锥形。叶多三角形，少量为扇形叶，无明显中裂。种实宽卵圆形，先端宽秃、顶部钝圆，呈"O"字形凹入，珠孔孔迹明显，具小尖。基部稍狭，蒂盘正圆，表面与周缘凹凸不平，周缘凹陷，但较浅。果柄长2.95 cm，弯曲。种实大小为纵径2.84 cm，横径2.81 cm，熟时绿黄色，被薄白粉，具透明梭状油胞。种核卵圆形，先端圆钝，具突尖，珠孔迹明显，基部束迹迹点特小，仅隐约可见，相距1.6~2.8 mm，分处于尾尖两侧。两侧棱线明显，但自中部稍下即不清晰。种核大小为2.25 cm×1.50 cm×1.22 cm，千粒重2 230 g。出核率27.42%，出仁率77.12%。广西桂林灵川海洋乡、山东海阳朱吴乡均见分布。

'Qingpi Guo'

RLW 1.50. Crown conoid. Leaves triangle or fan-shaped, without conspicuous 2-lobed. Fruits ovoid, blunt and broad with conspicuous micropylar and cusps at the apex, blunt and round at the tip, sunk in O-form. The base of stalk round, surface scraggly, and circumferentia slight concave; stalk slight oblique, 2.95 cm long; fruits orange yellow and covered with thin glaicous when ripe. Nuts asymmetrical, ovoid, broadly round and emarginated, with cusps and conspicuous micropylar at the apex, with two little vascular bundle scars at the base with distance 1.6–2.8 mm. Raphe conspicuous from the midst to the apex, nuts with pinhole-shaped. Size of nut is 2.25 cm×1.50 cm×1.22 cm. 1 000-grain weight of 2 230 g. PNF 27.42% and PKN 77.12% normally. Mainly cultivated in Haiyang of Lingchuan, Guilin, Guangxi and Zhuwu of Haiyang, Yantai, Shandong.

图 3.43
'青皮果'（原株）
①广西桂林海洋乡40年生实生大树；
②结果枝生长状态

Fig. 3.43
'Qingpi Guo'
① 40 years old tree from seedling in Guilin, Guangxi，
② Bearing shoots

银杏核用品种——佛指品种群
Fozhi Group for Nut-producing

图 3.44 '青皮果'
① 种实和种核；
② 种实和叶子；
③ 结果枝蜡叶标本

Fig. 3.44 'Qingpi Guo'
① Fruits and nuts,
② Fruits and leaves,
③ Wax-leaf specimens of bearing shoots

3 '鸭尾银杏'

又称'鸭屁股银杏',种核顶部具小尖常凹入,似鸭尾状,故名。核形系数1.56。树冠为自然开心形或圆头形,无中心主干。叶片中厚,叶面多皱摺。叶缘锯齿大而圆,无深裂。种实长圆形,中部稍大,顶部狭圆,果肩小而平,不歪斜。蒂盘椭圆或长圆,平或微凸。熟时橘黄色,被白粉。早熟品种。果柄平均长3.51 cm,自基部至先端渐细,基部粗0.32 (0.30～0.38) cm。种实大小为纵径3.11 cm,横径2.80 cm。种核广卵圆形,先端宽圆,顶部具小尖常凹入,似鸭尾状。种核基部狭圆,两束迹迹点明显,大而凸出。两侧棱狭而薄。种核之大小为2.37 cm×1.50 cm×1.26 cm,千粒重2 280 g。出核率20%,出仁率80.18%。位于江苏吴县东山镇。

'Yawei Yinxing'

'Yapigu Yinxing'.

Nuts duck tail-shaped. RLW 1.56. Crown spheroidal or open center-shaped. Leaf fan-shaped, thicker, covered with plicature, margin sinuous. Fruits long spheroidicity, blunt and round at the apex, the base of stalk long round or elliptical, exasperate, orange yellow, covered with glaicous when ripe; stalk 3.51cm long. Nuts broad ovoid, symmetrical, round and wider, with foveolate cusps at the apex, nut with conspicuous vascular bundle scars at the base, raphe narrow and thin. Size of nut is 2.37 cm×1.50 cm×1.26 cm. 1 000-grain weight of 2 280 g. PNF 20% and PKN 80.18%. Mainly cultivated in Dongshan of Wuxian, Jiangsu.

图 3.45 '鸭尾银杏'
① 江苏吴县实生树;
② 结果枝生长状态

Fig. 3.45 'Yawei Yinxing'
① Tree from seedling in Wuxian, Jiangsu
② Bearing shoots

银杏核用品种——佛指品种群
Fozhi Group for Nut-producing

图 3.46
'鸭尾银杏'
① 种实和种核；
② 种实和叶子；
③ 结果枝蜡叶标本

Fig. 3.46
'Yawei Yinxing'
① Fruits and nuts,
② Fruits and leaves,
③ Wax-leaf specimens of bearing shoots

4 '长糯白果'

核形系数为1.58。树冠一般呈塔形或圆头形，树皮灰褐色，有纵裂。叶色深绿，宽3～7.7 cm，长为4～7 cm。叶柄长3.0～7.7 cm。种实长卵形，果顶略钝，基部平阔，稍歪向一边。珠托中大，不规则，向一边倾斜，表面隆起，边缘不整齐。果面黄橙色，有果粉。种核长卵形，先端宽圆，顶尖凹陷，两维管束迹迹点小而明显，两束迹相距较宽的可达0.41 cm，但亦见二迹点合为一体者。种实下半部粗糙并有窄棱边，上半部略宽于下半部。种核较大，壳白色。种核之大小为2.24 cm×1.42 cm×1.20 cm。出核率18%。产于贵州盘县。

'Changnuo Baiguo'

RLW 1.58. Crown tower-shaped or spheroidal. Bark taupe with longitudinal fissures. Leaves deep green, 4–7 cm long, 3.0–7.7cm wide, petiole 3.0–7.7 cm long. Fruits long ovoid, slight obtuse at the apex, flattened and slight oblique at the base, ovule bracket irregular, gibbous, oblique, margin irregular, orange yellow and glaicous when ripe. Nuts larger, long ovoid, broadly rounded and emarginated at the apex, conspicuous and smaller vascular bundle points with broad distance or adnate, muricate with narrowly attenuate raphe at the base, sclerotesta white. Size of nut is 2.24 cm×1.42 cm×1.20 cm. PNF 18% normally. The tree is native to Panxian, Guizhou.

图 3.47
'长糯白果'
①江苏邳州陈楼银杏种质基因库嫁接树；
②结果枝生长状态

Fig. 3.47
'Changnuo Baiguo'
① Grafted tree in Ginkgo resources gene bank in Pizhou, Jiangsu,
② Bearing shoots

银杏核用品种——佛指品种群
Fozhi Group for Nut-producing

图 3.48
'长糯白果'
①种实和种核；
②种实和叶子；
③结果枝蜡叶标本

Fig. 3.48
'Changnuo Baiguo'
① Fruits and nuts,
② Fruits and leaves,
③ Wax-leaf specimens of bearing shoots

5 '家佛指'

又名'泰兴大白果'或'佛指'。核形系数1.63。叶片一般无明显缺刻,幼树叶大而肥厚。叶长×宽为8.57 cm×9.83 cm,叶柄长3.17 cm,佛指种实为长卵圆形,因种核似佛像手指而得名。珠托歪,边缘亦凸凹不平,种实歪托,种实略有弯曲,顶端圆钝孔迹小,平或稍下凹,但亦有少数或微小尖凸起的。成熟时外种皮橙黄色、偏淡,外被薄层白粉。种核长卵圆形,尾端长细圆秃,前端短圆钝;种核两侧有棱,无翼状,近尾端棱不明显,两束迹小而合二为一;壳洁白。种核之大小为2.50 cm×1.54 cm×1.45 cm。千粒重2 941 g。出核率26%~27%。种壳较薄,出仁率达78%~83%,一般80%。本品种为江苏泰兴银杏种子产区的主栽品种。

'Jia Fozhi'

'Taixing Dabaiguo,' 'Fozhi'.

RLW 1.63. Leaves not conspicuous incised, long 8.57 cm, wide 9.83 cm, petioles long 3.17 cm. Fruits long ovoid, slight oblique, finger-shaped, ovule bracket oblique, margin emarginate, blunt or rounded at the apex, flattened or emarginate at the base, with cusps, micropylar smaller, orange yellow and glaicous when ripe. Nuts long ovoid, blunt or round at the apex, attenuate and round at the base, raphe not wing-shaped, sclerotesta white; vascular bundle scars adnate. Size of nut is 2.50 cm×1.54 cm×1.45 cm. 1 000-grain weight of 2 941 g. PNF 26%–27% and PKN 78%–83%. Mainly cultivated in Taixing, Jiangsu.

图 3.49
'家佛指'(原株)
①江苏泰兴张桥镇25年生实生大树;
②结果枝生长状态

Fig. 3.49
'Jia Fozhi'
① 25 years old tree from seedling in Taixing, Jiangsu, ② Bearing shoots

图 3.50
'家佛指'
① 种实和种核；
② 种实和叶子；
③ 结果枝蜡叶标本

Fig. 3.50
'Jia Fozhi'
① Fruits and nuts,
② Fruits and leaves,
③ Wax-leaf specimens of bearing shoots

6 '洞庭皇'

核形系数1.56。本品种大树树冠多圆头形。皮灰褐色，叶大多为扇形，叶淡绿色，中裂浅，叶波状明显，叶片平展。种实长圆形、广卵圆形或倒卵圆形，丰满。果先端钝圆。珠孔小而长，鱼嘴状，缘有浅缺裂，高低不等，边缘有浅紫色线状条纹；珠孔迹小而明显。基部平而微凹，略有隆起，并向一侧歪斜。蒂盘长圆形或椭圆形，微突，表面高低不平。熟时淡橘黄色，被较厚白粉，有淡褐色油胞。核卵状长椭圆形，无脊腹之分。先端宽圆渐尖，具小突尖，中部以下渐狭，基部广楔形，近基部稍显粗糙。束迹迹点宽短，少数不显，迹点一高一低呈倾斜截面。两侧棱线宽窄不一，但无翼状边缘，长短也不相等。种核之大小为3.10 cm×2.00 cm×1.50 cm。千粒重3 597 g，出核率23%，出仁率78%。

'Dongting Huang'

RLW1.56. Crown spheroidal. Bark brownish-grey. Leaves fan-shaped, 2-lobed, margin sinuous, explanate, greenish. Fruits long spherical, ovoid or obovoid, blunt and round at the apex, micropylar smaller and longer, fish mouth-like, margin slight incised and with lavender stria; micropylar scar smaller, conspicuous, flattened and emarginate, slight protuberant, oblique at the base; the base of stalk long round or elliptical, exasperate, orange yellow, with glaicous and hazelly ocellus when ripe. Nuts long ellipsoid, symmetrical, rounded and acuminate at the apex with cusps, broad cuneate and muricate at the base, vascular bundle points broader and shorter, raphe asymmetrical, margin not wing-shaped. Size of nut is 3.10 cm×2.00 cm×1.50 cm. 1 000-grain weight of 3 597 g. PNF 23% and PKN 78%.

图3.51 '洞庭皇'
① 江苏邳州陈楼银杏种质基因库结果枝嫁接树；
② 结果枝生长状态

Fig. 3.51 'Dongting Huang'
① grafted tree in Ginkgo resources gene bank in Pizhou, Jiangsu,
② Bearing shoots

银杏核用品种——佛指品种群
Fozhi Group for Nut-producing

图 3.52 '洞庭皇'
① 种实和种核；
② 种实和叶子；
③ 结果枝蜡叶标本

Fig. 3.52
'Dongting Huang'
① Fruits and nuts,
② Fruits and leaves,
③ Wax-leaf specimens of bearing shoots

7 '郯魁'

又名'郯306号'，核形系数1.50。母树在港上镇王桥村，树龄90年，树势健旺，丰产性能好。叶片浓绿且肥厚，叶缺裂较深。种实千粒重13.42 kg；种核为倒卵形，千粒重3 560 g，出核率27%，出仁率78%。

'Tan Kui'

'Tan-306'.
RLW1.50.The tree 90 years old, grows in Gangshang of Pizhou, Jiangsu. Leaves deep 2-lobed, larger. Fruits 1000-grain of weight of 13.42 kg. Nuts obovoid. 1 000-grain weight of 3 560 g, PNF 27% and PKN 78%.

图 3.53 '郯魁' 结果枝生长状态

Fig. 3.53 'Tan Kui' Bearing shoots

银杏核用品种——佛指品种群
Fozhi Group for Nut-producing

图 3.54 '郯魁'
① 种实和种核；
② 种实和叶子；
③ 结果枝蜡叶标本

Fig. 3.54 'Tan Kui'
① Fruits and nuts,
② Fruits and leaves,
③ Wax-leaf specimens of bearing shoots

8 '贵州长白果'

核形系数1.61。树冠呈塔形或长圆头形，高可达30 m。长枝上部叶片呈窄扇形，中下部叶为扇形，有浅裂至中裂。短枝叶扇形，有浅裂。叶宽4.5~7.0 cm，叶柄长6~7 cm。种实卵圆形，略偏斜，成熟时黄橙色，被白粉，表面粗糙，先端圆，顶尖凹下。珠托近圆形，中大，向一面歪斜，表面不平，边缘略凹下。种核长卵圆形，形状近似佛指。先端突尖，中部以下较窄，两束迹迹点小，两侧有棱线，中上部较明显。种核大小为2.4 cm×1.49 cm×1.20 cm，千粒重2 000 g，出核率19%，出仁率79%。产于贵州盘县。

'Guizhou Chang Baiguo'

RLW1.61. Crown tower-shaped or long spherical. Leaves narrow fan-shaped on upper of long branchlets, fan-shaped with fissures on beneath of long branchlets, fan-shaped with shallow fissures on short branchlets; petiole 6–7cm long. Fruits ellipsoid-spherical, oblique, muricate, round and emarginated at the apex, ovule bracket subspherical, middle size, oblique, margin emarginate, orange yellow and glaicous when ripe. Nuts long ellipsoid-spherical, Buddha's finger -shaped, with cusps at the apex, turning narrower beneath midst, vascular bundle points smaller, raphe conspicuous from the midst to the apex. Size of nut is 2.4 cm×1.49 cm×1.20 cm. 1 000-grain weight of 2 000 g. PNF 19% and PKN 79%. The tree is native to Panxian, Guizhou.

图 3.55
'贵州长白果'
① 贵州盘县结果枝嫁接树；
② 结果枝生长状态

Fig. 3.55 'Guizhou Chang Baiguo'
① Grafted tree in Panxian, Guizhou, ② Bearing shoots

银杏核用品种——佛指品种群
Fozhi Group for Nut-producing

图 3.56
'贵州长白果'
① 种实和种核；
② 种实和叶子；
③ 结果枝蜡叶标本

Fig. 3.56
'Guizhou Chang Baiguo'
① Fruits and nuts,
② Fruits and leaves,
③ Wax-leaf specimens of bearing shoots

9 '黄皮果'

核形系数1.53。多扇形叶,中裂明显,叶长4.7 cm,宽6.8 cm。种实卵圆形,先端浑圆,顶部宽大圆秃,具小尖,珠孔孔迹明显。种实有明显缢痕。蒂盘圆形,周缘整齐,略凹陷。果柄长4.0~4.8 cm,略弯曲。熟时鲜黄色,被薄白粉,外皮较薄,无透明枝状油胞,多双果。种实大小为纵径2.7 cm,横径2.3 cm。种核卵圆形,上下几乎相等,先端秃尖,珠孔孔迹明显。基部两束迹迹点大而长,相距1.5 mm,中间有木质化横隔相连。两侧棱线明显,但仅在上部可见,下部不显。种核大小为2.37 cm×1.55 cm×1.36 cm,种核千粒重2 390 g。出核率24.96%,出仁率76.22%。主要分布于广西灵川,浙江临安。

'Huangpi Guo'

RLW 1.53. Leaves fan-shaped, 4.7 cm long and 6.8 cm wide, with conspicuous 2-lobed. Fruits ovoid, with conspicuous constriction, blunt and broad with conspicuous micropylar and cusp at the apex. The base of stalk round and the circumferentia slight concave, stalk slight oblique, 4.0–4.8 cm long; fruits yellow and covered with thin glaicous, but without hyaloid oil sac when ripe. Nuts symmetrical, ovoid, with cusps and conspicuous micropylar at the apex, with big and adnate vascular bundle scars at the base with distance 1.5 mm. Raphe conspicuous above the midst. Size of nut is 2.37 cm×1.55 cm×1.36 cm. 1 000-grain weight of 2 390 g. PNF 24.96% and PKN 76.22% normally. Mainly cultivated in Lingchuan, Guangxi and Lin'an, Zhejiang.

图 3.57
'黄皮果'(原株)
①广西桂林海洋乡40年生实生大树;
②结果枝生长状态

Fig. 3.57
'Huangpi Guo'
① 40 years old tree from seedling in Guilin, Guangxi,
② Bearing shoots

银杏核用品种——佛指品种群
Fozhi Group for Nut-producing

图 3.58 '黄皮果'
① 种实和种核；
② 种实和叶子；
③ 结果枝蜡叶标本

Fig. 3.58 'Huangpi Guo'
① Fruits and nuts,
② Fruits and leaves,
③ Wax-leaf specimens of bearing shoots

10 '金坠 13 号'

核形系数1.50。该品种属于佛指类，母树在山东郯城县新村乡新一村，树龄150年。枝条灰白色、有弯曲，芽体小，椭圆形。叶片厚而浓绿。种核狭长，顶有尖，基部维管束合生，侧棱线下部2/5处不明显。千粒重2 600 g，出核率25%，出仁率78%。

'Jinzhui-13'

RLW 1.50. The tree grows in Xincun of Tancheng, Shandong, grafting tree, 150 years old. Branchlets pallid, slight oblique; buds ellipsoid, smaller; leaves larger, deep green. Nuts ellipsoid, with cusps at the apex, adnate vascular bundle scars at the base, raphe conspicuous above midst. 1 000-grain weight of 2 600 g. PNF 25% and PKN 78% normally.

图 3.59
'金坠 13 号'
结果枝生长状态

Fig. 3.59
'Jinzhui-13'
Bearing shoots

银杏核用品种——佛指品种群
Fozhi Group for Nut-producing

图 3.60 '金坠 13 号'
① 种实和种核；
② 种实和叶子；
③ 结果枝蜡叶标本

Fig. 3.60 'Jinzhui-13'
① Fruits and nuts,
② Fruits and leaves,
③ Wax-leaf specimens of bearing shoots

11 '华口大白果'

核形系数1.56。种实长椭圆形，种实先端圆弧形，顶点凹，略见珠孔迹，基部平，蒂盘呈圆形或椭圆形，果柄长2.8～3.7 cm，上粗下细，略有弯曲。种实成熟时为橙黄色，表皮有油胞，并覆盖一层白粉。种核卵形，核粒较大，中上部较宽，中下部狭窄，核粒先端圆钝，顶部无明显小尖，基部尖，略见两维管束迹，两迹点紧成直线，尖凸状。两侧棱较明显。种核大小为2.85 cm×1.82 cm×1.45 cm。千粒重3 900 g。出核率32%，出仁率78%。核仁黄白色。主产于广西灵川。

'Huakou Da Baiguo'

RLW 1.56. Fruits long ellipsoid, arc and emarginate at the apex, micropylar scar slight consipicuous, flattened at the base, the base of stalk rounded or elliptic, funiculus 2.8–3.7 cm long, oblique and acuminate, Fruits orange yellow with glaicous and ocellus when ripe. Nuts larger, ovoid, attenuate from apex to base, blunt or rounded at the apex without conspicuous cusp, inconspicuous adnate vascular bundle scars in a line with cusp at the base, raphe conspicuous. Kernels yellowish white. Size of nut is 2.85 cm×1.82 cm×1.45 cm. 1 000-grain weight of 3 900 g. PNF 32% and PKN 78% normally. Mainly cultivated in Lingchuan, Guangxi.

图 3.61
'华口大白果'
①广西桂林海洋乡嫁接树；
②结果树

Fig. 3.61
'Huakou Da Baiguo'
① Grafted tree in, Guilin, Guangxi,
② Bearing shoots

银杏核用品种——佛指品种群
Fozhi Group for Nut-producing

图 3.62 '华口大白果'
① 种实和种核；
② 种实和叶子；
③ 结果枝蜡叶标本

Fig. 3.62
'Huakou Da Baiguo'
① Fruits and nuts,
② Fruits and leaves,
③ Wax-leaf specimens of bearing shoots

12 '新宇'

又名'金坠1号'。核形系数1.69。叶长×宽为6.02 cm×8.38 cm，叶柄长6.28 cm，叶缘呈波浪状。种实倒卵形，种柄直立。核顶端有尖，基部两束迹合生，背腹明显。最大种核3.8 g，种壳厚0.38 mm。核长×宽×厚为2.5 cm×1.47 cm×1.34 cm。千粒重3 300 g，出核率28.14%，出仁率80.34%。

'Xin Yu'

'Jinzhui-1'.

RLW 1.69. Leaves margin repand, 6.02 cm long, 8.38 cm wide, petiole 6.28cm long. Fruits obovoid; stalk erect. Nuts with cusps at the apex, adnate vascular bundle scars at the base, sclerotesta 0.38 mm thick. Size of nut is 2.50 cm×1.47 cm×1.34 cm. 1 000-grain weight of 3 300 g. PNF 28.14% and PKN 80.34% normally.

图 3.63 '新宇'
山东郯城清泉寺林场银杏种质基因库结果枝生长状态

Fig. 3.63 'Xin Yu'
Bearing shoots of Grafted tree in Ginkgo resources gene bank in Tancheng, Shandong

银杏核用品种——佛指品种群
Fozhi Group for Nut-producing

图 3.64 '新宇'
① 种实和种核；
② 种实和叶子；
③ 结果枝蜡叶标本

Fig. 3.64 'Xin Yu'
① Fruits and nuts,
② Fruits and leaves,
③ Wax-leaf specimens of bearing shoots

13 '扁佛指'

本品种的种核近似佛指，但背厚腹薄，呈扁平状，故名。核形系数1.62。叶有多扇形及截形叶，明显中裂，叶长较厚，色较深。种实为不正广长圆形，亦略有偏斜，但正托较多。孔迹平或稍凹陷，少有孔迹呈尖状凸起的。种实长3.2～3.4 cm，宽2.2～2.4 cm。珠托较大，珠柄长5 cm许，比佛指明显短。熟时橙黄色，被厚白粉。种核背厚腹薄而略扁，下半部比上半部狭；两束迹明显大，相距亦较远，有时呈鸭尾形。种核壳比佛指稍厚；种仁亦常不如佛指饱满，摇之易有响声。种核长×宽×厚为3.0 cm×1.85 cm×1.35 cm。种核千粒重2 500 g，出核率为25%左右，出仁率为79%。

'Bian Fozhi'

Nuts Buddha's finger-shaped. RLW 1.62. Leaves fan-shaped and cuneate, conspicuous 2-lobed, longer and thicker, deep color. Fruits long spherical, erect usually, rarely oblique, long 3.2–3.4 cm, wide 2.2–2.4 cm, micropylar scar flattened or emarginated, rarely apiculate, ovule bracket larger; stalk 5 cm long, fruits orange yellow and cover with thick glaicous when ripe. Nuts asymmetrical, attenuate from the apex to the base, vascular bundle scars conspicuous, far apart, occasionally ducktail-shaped, sclerotesta thicker than Fozhi Group's, kernels nosing while shaking. Size of nut is 3.0 cm×1.85 cm×1.35 cm. 1 000-grain weight of 2 500 g, PNF 25% and PKN 79%.

图3.65 '扁佛指'
①江苏泰兴嫁接树；
②结果枝生长状态

Fig. 3.65 'Bian Fozhi'
① Grafted tree in Taixing, Jiangsu,
② Bearing shoots

银杏核用品种——佛指品种群
Fozhi Group for Nut-producing

图 3.66 '扁佛指'
① 种实和种核；
② 种实和叶子；
③ 结果枝蜡叶标本

Fig. 3.66
'Bian Fozhi'
① Fruits and nuts,
② Fruits and leaves,
③ Wax-leaf specimens of bearing shoots

3.1.2.3 佛指品种群优良无性系

本节分别对佛指品种群中 25 个优良无性系进行简要介绍，见表 3.2。

3.1.2.3 Clones of Fozhi Group

Table 3.2 Illustrated 25 clones of Fozhi group in brief.

表 3.2 佛指品种群优良无性系

Table 3.2 A list of Some Ginkgo Clones of Fozhi Group

编号 Number	名称 Name	核形系数 RLW	产地 Distribution
1	'桂林 9 号' 'Guilin-9'	1.50	广西桂林 Guilin, Guangxi
2	'叶籽银杏' 'Ye Zi Yinxing'	1.51	山东沂源 Yiyuan, Shandong
3	'京山 A25 号' 'Jingshan-A25'	1.52	湖北京山 Jingshan, Hubei
4	'郯新' 'Tan Xin'	1.52	山东郯城新村 Xincun, Tancheng, Shandong
5	'曹 2 号' 'Cao-2'	1.55	山东港上曹楼 Caolou, Gangshang, Shandong
6	'曹 1 号' 'Cao-1'	1.56	山东港上曹楼 Caolou, Gangshang, Shandong
7	'新村 202 号' 'Xincun-202'	1.56	山东郯城新村 Xincun, Tancheng, Shandong
8	'重坊 106 号' 'Chongfang-106'	1.56	山东郯城重坊 Chongfang, Tancheng, Shandong
9	'港西 2 号' 'Gangxi-2'	1.561	山东港上曹楼 Caolou, Gangshang, Shandong
10	'泰兴 3 号' 'Taixing-3'	1.60	扬州大学 Yangzhou University
11	'长兴 5 号' 'Changxing-5'	1.61	浙江长兴 Changxing, Zhejiang
12	'泰兴 1 号' 'Taixing-1'	1.61	扬州大学 Yangzhou University
13	'新村 9 号' 'Xincun-9'	1.616	山东郯城新村 Xincun, Tancheng, Shandong
14	'重坊 111 号' 'Chongfang-111'	1.638	山东郯城重坊 Chongfang, Tancheng, Shandong

续表

编号 Number	名称 Name	核形系数 RLW	产地 Distribution
15	'郯城231号' 'Tancheng-231'	1.64	山东郯城县 Tancheng, Shandong
16	'港上303号' 'Gangshang-303'	1.65	山东郯城港上 Gangshang, Tancheng, Shandong
17	'新村210号' 'Xincun-210'	1.65	山东郯城新村 Xincun, Tancheng, Shandong
18	'洞庭佛手1号' 'Dongting Foshou-1'	1.66	扬州大学 Yangzhou University
19	'苏农佛手' 'Sunong Foshou'	1.66	扬州大学 Yangzhou University
20	'新村401号' 'Xincun-401'	1.66	山东郯城新村 Xincun, Tancheng, Shandong
21	'正安1号' 'Zheng'an-1'	1.66	贵州正安 Zheng'an, Guizhou
22	'长兴2号' 'Changxing-2'	1.67	浙江长兴 Changxing, Zhejiang
23	'港上501号' 'Gangshang-501'	1.69	山东郯城港上 Gangshang, Tancheng, Shandong
24	'泰兴2号' 'Taixing-2'	1.694	江苏泰兴 Taixing, Jiangsu
25	'古银杏' 'Gu Yinxing'	1.699	山东莒县定灵寺 Dingling Temple, Juxian, Shandong

'桂林 9 号' 'Guilin-9'

图 3.67
'桂林 9 号'
结果枝生长状态

Fig. 3.67
'Guilin-9'
Bearing shoots

图 3.68
'桂林 9 号'
① 结果枝蜡叶标本；
② 种实和种核；
③ 种实和叶子

Fig. 3.68
'Guilin-9'
① Wax-leaf specimens of bearing shoots,
② Fruits and nuts,
③ Fruits and leaves

银杏核用品种——佛指品种群
Fozhi Group for Nut-producing

'叶籽银杏' 'Ye Zi Yinxing'

图 3.69
'叶籽银杏'
结果枝生长状态

Fig. 3.69
'Ye Zi Yinxing'
Bearing shoots

图 3.70
'叶籽银杏'
① 种实和种核；
② 种实和叶子；
③ 结果枝蜡叶标本

Fig. 3.70
'Ye Zi Yinxing'
① Fruits and nuts,
② Fruits and leaves,
③ Wax-leaf specimens of bearing shoots

第3章 银杏品种检索表及品种图谱
Chapter 3　Key to Cultivars of *Ginkgo biloba* L. and Illustration of Ginkgo Cultivars

'京山 A25 号' 'Jingshan-A25'

图 3.71
'京山 A25 号'
结果枝生长状态

Fig. 3.71
'Jingshan-A25'
Bearing shoots

图 3.72
'京山 A25 号'
①结果枝蜡叶标本；
②种实和种核；
③种实和叶子

Fig. 3.72
'Jingshan-A25'
① Wax-leaf specimens of bearing shoots,
② Fruits and nuts,
③ Fruits and leaves

'郯新' 'Tan Xin'

图 3.73
'郯新'
结果枝生长状态

Fig. 3.73
'Tan Xin'
Bearing shoots

图 3.74
'郯新'
①种实和种核；
②种实和叶子；
③结果枝蜡叶标本

Fig. 3.74
'Tan Xin'
① Fruits and nuts,
② Fruits and leaves,
③ Wax-leaf specimens of bearing shoots

'曹2号' 'Cao-2'

图 3.75
'曹 2 号'
结果枝生长状态

Fig. 3.75
'Cao-2'
Bearing shoots

图 3.76
'曹 2 号'
①结果枝蜡叶标本；
②种实和种核；
③种实和叶子

Fig. 3.76
'Cao-2'
① Wax-leaf specimens of bearing shoots,
② Fruits and nuts,
③ Fruits and leaves

银杏核用品种——佛指品种群
Fozhi Group for Nut-producing

'曹1号' 'Cao-1'

图 3.77
'曹1号'
结果枝生长状态

Fig. 3.77
'Cao-1'
Bearing shoots

图 3.78
'曹1号'
①种实和种核；
②种实和叶子；
③结果枝蜡叶标本

Fig. 3.78
'Cao-1'
① Fruits and nuts,
② Fruits and leaves,
③ Wax-leaf specimens of bearing shoots

第3章 银杏品种检索表及品种图谱
Chapter 3　Key to Cultivars of *Ginkgo biloba* L. and Illustration of Ginkgo Cultivars

'新村202号'　'Xincun-202'

图 3.79
'新村 202 号'
结果枝生长状态

Fig. 3.79
'Xincun-202'
Bearing shoots

图 3.80
'新村 202 号'
①结果枝蜡叶标本；
②种实和种核；
③种实和叶子

Fig. 3.80
'Xincun-202'
① Wax-leaf specimens of bearing shoots,
② Fruits and nuts,
③ Fruits and leaves

86

银杏核用品种——佛指品种群
Fozhi Group for Nut-producing

'重坊 106 号' 'Chongfang-106'

图 3.81
'重坊 106 号'
结果枝生长状态

Fig. 3.81
'Chongfang-106'
Bearing shoots

图 3.82
'重坊 106 号'
① 种实和种核；
② 种实和叶子；
③ 结果枝蜡叶标本

Fig. 3.82
'Chongfang-106'
① Fruits and nuts,
② Fruits and leaves,
③ Wax-leaf specimens of bearing shoots

87

第 3 章 银杏品种检索表及品种图谱
Chapter 3　Key to Cultivars of *Ginkgo biloba* L. and Illustration of Ginkgo Cultivars

'港西 2 号' 'Gangxi-2'

图 3.83
'港西 2 号'
结果枝生长状态

Fig. 3.83
'Gangxi-2'
Bearing shoots

图 3.84
'港西 2 号'
① 结果枝蜡叶标本；
② 种实和种核；
③ 种实和叶子

Fig. 3.84
'Gangxi-2'
① Wax-leaf specimens of bearing shoots,
② Fruits and nuts,
③ Fruits and leaves

'泰兴3号' 'Taixing-3'

图 3.85
'泰兴3号'
结果枝生长状态

Fig. 3.85
'Taixing-3'
Bearing shoots

图 3.86
'泰兴3号'
① 种实和种核;
② 种实和叶子;
③ 结果枝蜡叶标本

Fig. 3.86
'Taixing-3'
① Fruits and nuts,
② Fruits and leaves,
③ Wax-leaf specimens of bearing shoots

'长兴5号' 'Changxing-5'

图 3.87
'长兴5号'
结果枝生长状态
Fig. 3.87
'Changxing-5'
Bearing shoots

图 3.88
'长兴5号'
①结果枝蜡叶标本；
②种实和种核；
③种实和叶子
Fig. 3.88
'Changxing-5'
① Wax-leaf specimens of bearing shoots,
② Fruits and nuts,
③ Fruits and leaves

'泰兴1号' 'Taixing-1'

图 3.89
'泰兴1号'
结果枝生长状态

Fig. 3.89
'Taixing-1'
Bearing shoots

图 3.90
'泰兴1号'
① 种实和种核；
② 种实和叶子；
③ 结果枝蜡叶标本

Fig. 3.90
'Taixing-1'
① Fruits and nuts,
② Fruits and leaves,
③ Wax-leaf specimens of bearing shoots

'新村9号' 'Xincun-9'

图 3.91
'新村9号'
结果枝生长状态

Fig. 3.91
'Xincun-9'
Bearing shoots

图 3.92
'新村9号'
①结果枝蜡叶标本；
②种实和种核；
③种实和叶子

Fig. 3.92
'Xincun-9'
① Wax-leaf specimens of bearing shoots,
② Fruits and nuts,
③ Fruits and leaves

'重坊111号' 'Chongfang-111'

图 3.93
'重坊111号'
结果枝生长状态

Fig. 3.93
'Chongfang-111'
Bearing shoots

图 3.94
'重坊111号'
①种实和种核；
②种实和叶子；
③结果枝蜡叶标本

Fig. 3.94
'Chongfang-111'
① Fruits and nuts,
② Fruits and leaves,
③ Wax-leaf specimens of bearing shoots

'郯城231号' 'Tancheng-231'

图 3.95
'郯城231号'
结果枝生长状态

Fig. 3.95
'Tancheng-231'
Bearing shoots

图 3.96
'郯城231号'
①结果枝蜡叶标本；
②种实和种核；
③种实和叶子

Fig. 3.96
'Tancheng-231'
① Wax-leaf specimens of bearing shoots,
② Fruits and nuts,
③ Fruits and leaves

银杏核用品种——佛指品种群
Fozhi Group for Nut-producing

'港上303号' 'Gangshang-303'

图 3.97
'港上303号'
结果枝生长状态

Fig. 3.97
'Gangshang-303'
Bearing shoots

图 3.98
'港上303号'
①种实和种核；
②种实和叶子；
③结果枝蜡叶标本

Fig. 3.98
'Gangshang-303'
① Fruits and nuts,
② Fruits and leaves,
③ Wax-leaf specimens of bearing shoots

'新村210号' 'Xincun-210'

图 3.99
'新村 210 号'
结果枝生长状态

Fig. 3.99
'Xincun-210'
Bearing shoots

图 3.100
'新村 210 号'
① 结果枝蜡叶标本；
② 种实和种核；
③ 种实和叶子

Fig. 3.100
'Xincun-210'
① Wax-leaf specimens of bearing shoots,
② Fruits and nuts,
③ Fruits and leaves

'洞庭佛手1号' 'Dongting Foshou-1'

图 3.101
'洞庭佛手1号'
结果枝生长状态

Fig. 3.101
'Dongting Foshou-1'
Bearing shoots

图 3.102
'洞庭佛手1号'
① 种实和种核；
② 种实和叶子；
③ 结果枝蜡叶标本

Fig. 3.102
'Dongting Foshou-1'
① Fruits and nuts,
② Fruits and leaves,
③ Wax-leaf specimens of bearing shoots

第3章 银杏品种检索表及品种图谱
Chapter 3　Key to Cultivars of *Ginkgo biloba* L. and Illustration of Ginkgo Cultivars

'苏农佛手'　'Sunong Foshou'

图 3.103
'苏农佛手'
结果枝生长状态

Fig. 3.103
'Sunong Foshou'
Bearing shoots

图 3.104
'苏农佛手'
①结果枝蜡叶标本；
②种实和种核；
③种实和叶子

Fig. 3.104
'Sunong Foshou'
① Wax-leaf specimens of bearing shoots,
② Fruits and nuts,
③ Fruits and leaves

'新村401号' 'Xincun-401'

图 3.105
'新村401号'
结果枝生长状态

Fig. 3.105
'Xincun-401'
Bearing shoots

图 3.106
'新村401号'
①种实和种核；
②种实和叶子；
③结果枝蜡叶标本

Fig. 3.106
'Xincun-401'
① Fruits and nuts,
② Fruits and leaves,
③ Wax-leaf specimens of bearing shoots

'正安1号' 'Zheng'an-1'

图 3.107
'正安1号'
结果枝生长状态

Fig. 3.107
'Zheng'an-1'
Bearing shoots

图 3.108
'正安1号'
① 结果枝蜡叶标本；
② 种实和种核；
③ 种实和叶子

Fig. 3.108
'Zheng'an-1'
① Wax-leaf specimens of bearing shoots,
② Fruits and nuts,
③ Fruits and leaves

银杏核用品种——佛指品种群
Fozhi Group for Nut-producing

'长兴 2 号' 'Changxing-2'

图 3.109
'长兴 2 号'
结果枝生长状态

Fig. 3.109
'Changxing-2'
Bearing shoots

图 3.110
'长兴 2 号'
① 种实和种核；
② 种实和叶子；
③ 结果枝蜡叶标本

Fig. 3.110
'Changxing-2'
① Fruits and nuts,
② Fruits and leaves,
③ Wax-leaf specimens of bearing shoots

第3章 银杏品种检索表及品种图谱
Chapter 3 Key to Cultivars of *Ginkgo biloba* L. and Illustration of Ginkgo Cultivars

'港上501号' 'Gangshang-501'

图 3.111
'港上 501 号'
结果枝生长状态

Fig. 3.111
'Gangshang-501'
Bearing shoots

图 3.112
'港上 501 号'
① 结果枝蜡叶标本；
② 种实和种核；
③ 种实和叶子

Fig. 3.112
'Gangshang-501'
① Wax-leaf specimens of bearing shoots,
② Fruits and nuts,
③ Fruits and leaves

银杏核用品种——佛指品种群
Fozhi Group for Nut-producing

'泰兴2号' 'Taixing-2'

图 3.113
'泰兴2号'
结果枝生长状态

Fig. 3.113
'Taixing-2'
Bearing shoots

图 3.114
'泰兴2号'
①种实和种核；
②种实和叶子；
③结果枝蜡叶标本

Fig. 3.114
'Taixing-2'
① Fruits and nuts,
② Fruits and leaves,
③ Wax-leaf specimens of bearing shoots

'古银杏' 'Gu Yinxing'

图 3.115
'古银杏'
结果枝生长状态

Fig. 3.115
'Gu Yinxing'
Bearing shoots

图 3.116
'古银杏'
①结果枝蜡叶标本；
②种实和种核；
③种实和叶子

Fig. 3.116
'Gu Yinxing'
① Wax-leaf specimens, of bearing shoots
② Fruits and nuts,
③ Fruits and leaves

3.1.3 中子品种群

将银杏种核长与宽的比值介于1.30（±0.05）与1.50（±0.05）之间的银杏品种确定为中子品种群，其中种核纵横轴线交叉点位于纵径的2/5处为马铃亚品种群；种核纵横轴线交叉点位于纵径的1/2处为梅核亚品种群，共包含了14个优良品种及14个优良无性系。本节重点介绍这14个优良品种特性，另有14个优良无性系只以表格和照片的形式做简要介绍。

3.1.3 Zhongzi Group

There are 14 cultivars and 14 clones of Zhongzi group, according to the ratio of length to width, range from 1.30(±0.05) to1.50(±0.05). In this group, the cultivars were classified into Maling Sub-group while the intersection of ordinate axis and horizontal axis of nut site two fifth of ordinate axis, and if the intersection of ordinate axis and horizontal axis of nut situates in one half of ordinate axis, they were classified into Meihe Sub-group.

广西桂林灵川古银杏群落

The Ancient Ginkgo Trees in Lingchuan, Guilin, Guangxi

3.1.3.1　中子品种群分类检索表

中子品种群检索表

1. 种核纵横轴线交叉点位于纵径的2/5处　　　　　　　　　　　　　　　　　　　　（马铃亚品种群）
　　2. 种核两面不对称，背圆腹薄
　　　　3. 种核中部有一条中缢隐线，侧棱明显有翼..1.'亚甜'
　　　　3. 种核中部无中缢隐线，两侧棱无翼..2.'李子果'
　　2. 种核两面对称，无背腹之分
　　　　4. 种核两面不具麻点
　　　　　　5. 种核中部有一条中缢隐线
　　　　　　　　6. 种实先端较圆
　　　　　　　　　　7. 种实基部正托，两维管束迹不歪...3.'魁金'
　　　　　　　　　　7. 种实基部非正托，两维管束迹稍歪..4.'魁铃'
　　　　　　　　6. 种实先端明显平阔，呈圆底状...5.'圆底果'
　　　　　　5. 种核中部无中缢隐线
　　　　　　　　8. 种核侧棱仅中上部较明显，不具翼状边缘
　　　　　　　　　　9. 种核基部两维管束迹点明显，不连成短横线
　　　　　　　　　　　　10. 种柄细，平直，无弯曲..6.'桂049号'
　　　　　　　　　　　　10. 种柄弯曲带钩..7.'观音皇'
　　　　　　　　　　9. 种核基部两维管束迹点相连成短横线..8.'海洋皇'
　　　　　　　　8. 种核侧棱明显，中部最宽处有翼状边缘...9.'宇香'
　　　　4. 种核两面具3~7个小麻点..10.'马铃5号'
1. 种核纵横轴线交叉点位于纵径的1/2处　　　　　　　　　　　　　　　　　　　　（梅核亚品种群）
　　11. 种核两面不对称，背圆腹薄
　　　　12. 两侧棱明显，中上部成窄翼状边缘..11.'梅核'
　　　　12. 上下均明显，无翼状边缘...12.'棉花果'
　　11. 种核两面对称，无背腹之分
　　　　13. 种核两侧棱线在1/2处消失...13.'早实梅核'
　　　　13. 种核两侧棱线（有的有翼）至基部4/5消..14.'安陆1号'

3.1.3.1 Key to Zhongzi Group

Key to Zhongzi Group of *Ginkgo biloba* L.

1. The intersection of ordinate axis and horizontal axis of nut situates in two fifth of ordinate axis **(Maling Sub-group)**
   - 2. Nuts asymmetrical
     - 3. Nuts with constriction at the midst, raphe wing-shaped.......................... 1. 'Ya Tian'
     - 3. Nuts without constriction at the midst, raphe not wing-shaped............. 2. 'Lizi Guo'
   - 2. Nuts symmetrical
     - 4. Nuts without foveolate pock
       - 5. Nuts with constriction at midst
         - 6. Fruits round at the apex
           - 7. The base of stalk and vascular bundle scars erect at the base...... 3. 'Kui Jin'
           - 7. The base of stalk and vascular bundle scars slight oblique at the base .... 4. 'Kui Ling'
         - 6. Fruits round bottomed, broad and smooth at the apex............ 5. 'Yuandi Guo'
       - 5. Nuts without constriction at the midst
         - 8. Raphe not wing-shaped, conspicuous from the midst to the apex
           - 9. Nuts with adnate vascular bundle scars but not strigula-shaped at the base
             - 10. Stalks fine, straight and erect ....6. 'Gui-049'
             - 10. Stalks oblique and with a little hook at the tip..... 7. 'Guanyin Huang'
           - 9. Nuts with adnate and strigula-shaped vascular bundle scars at the base... 8. 'Haiyang Huang'
         - 8. Nuts with conspicuous raphe and raphe wing-shaped at the midst ...... 9. 'Yuxiang'
     - 4. Nuts with foveolate pock range from 3 to 7... 10. 'Maling-5'
1. The intersection of ordinate axis and horizontal axis of nut sites at one half of ordinate axis **(Meihe Sub-group)**
   - 11. Nuts asymmetrical
     - 12. Nuts with conspicuous raphe and raphe wing-shaped from the midst to the apex .... 11. 'Meihe'
     - 12. Nuts with conspicuous not wing-shaped raphe .... 12. 'Mianhua Guo'
   - 11. Nuts symmetrical
     - 13. Nuts with raphe from the midst to the apex........................ 13. 'Zaoshi Meihe'
     - 13. Nuts with raphe from one fifth of the base to the apex .... 14. 'Anlu-1'

3.1.3.2 中子品种群优良品种

分别对中子品种群 14 个优良品种进行详细介绍。

3.1.3.2 Cultivars of Zhongzi Group

Illustrated 14 cultivars of Zhongzi Group in detail.

1 '亚甜'

又名'邳县大马铃'。核形系数1.41。圆柱形或卵圆形树冠。主干通直圆满,树势较强。叶片多为扇形,叶缘有多个深浅不等圆缺,有明显中裂。种实为广短卵圆形,似马铃形。前端大而圆秃,尾端如马铃,后半部缩小。上下之间似有很明显的中缢。孔迹平或微凹或有小尖。珠托圆形,多正托,边缘稍陷入种实。种柄直而宽扁。外种皮橙黄色,外被白粉。种核为短广卵圆形,前端宽大而圆,尾端向内收缩似圆形的下端。在种壳中间有一圈不很明显的隐约中缢横线。背圆浑而厚,腹略薄,故视之略扁。孔迹突出成尖,种核尾端两束迹相距较远。两侧有棱,明显有翼。种核大小为2.70 cm×1.91 cm×1.60 cm。千粒重是2 870 g。出核率29%,出仁率80%。主要分布于山东、江苏、浙江等。

'Ya Tian'

'Pixian Da Maling'.

RLW 1.41. Trunks straight and full. Crown cylindrical or ovoid. Leaves fan-shaped, margin emarginated, conspicuous 2-lobed. Fruits broad and short ovoid, horse bell-like, spheroidal and bald at the apex, horse bell-shaped at the base; micropylar flattened or emarginate or with cusps, ovule bracket spherical, erect, slightly sink into seed, stalk erect and plano-compressed, orange yellow and glaicous. Nuts broad and short ovoid, asymmetrical, larger and broad spherical at the apex, acuminate at the base; a inconspicuous annular crest at middle of nut, micropylar scar apiculate, vascular bundle sacrs far apart at the base, raphe wing-shaped. Size of nut is 2.70 cm×1.91 cm×1.60 cm. 1 000-grain weight of 2 870 g. PNF 29% and PKN 80%. Mainly cultivated in Shandong, Jiangsu and Zhejiang.

图 3.117 '亚甜'
①江苏邳州嫁接树;
②结果枝生长状态

Fig. 3.117 'Ya Tian'
① Grafted tree in Pizhou, Jiangsu,
② Bearing shoots

银杏核用品种——中子品种群
Zhongzi Group for Nut-producing

图 3.118 '亚甜'
① 种实和种核；
② 种实和叶子；
③ 结果枝蜡叶标本

Fig. 3.118 'Ya Tian'
① Fruits and nuts,
② Fruits and leaves,
③ Wax-leaf specimens of bearing shoots

2 '李子果'

本品种种实大小及形状均似蔷薇科之李子的种实，故名。核形指数1.43。树冠圆锥形，主干挺拔，层性明显。长枝上多三角形叶，少数为扇形叶，叶无明显中裂。种实卵圆形，熟时青黄色，被厚白粉，先端圆秃，顶微凹呈"一"字形，珠孔迹迹点明显，基部平。蒂盘椭圆形，略偏斜，周缘不整，略凹陷。种实柄直而粗壮，长约4.4 cm。种实大小为纵径2.51 cm，横径2.12 cm，单粒种实重6.4 g，每千克粒数156粒。多单果，少数双果。种核倒卵圆形，先端秃圆，顶部不具小尖。自中部向下呈圆锥形，基部两维管束迹迹点明显，分列尾尖两侧，相距1.7～2.7 mm。两侧棱自上至下均见，近尾部处不够清晰。种核有背腹之分，背厚腹薄略较明显。种核大小为2.0 cm×1.4 cm×1.05 cm，种核千粒重1 400～1 500 g。出核率25.21%，出仁率79.8%。主要分布于广西灵川、浙江诸暨等。

'Lizi Guo'

Fruits plum-shaped. RLW 1.43. Crown conoid, main trunk straight. Leaves triangle on long branchlets, without conspicuous 2-lobed. Fruits ovoid, foveolate, blunt and broad with conspicuous micropylar and cusps at the apex. The base of stalk elliptic, the surface scraggly, and the circumferentia slight concave. Stalks average 4.4 cm long. Fruits green yellow and covered with thicker glaicous when ripe, fruits 2.51 cm long, 2.12 cm wide. Nuts broad obovoid, the apex round and without cusp, the base with two conspicuous vascular bundle scars with the average distance 1.7–2.7 mm, nut asymmetrical. Size of nut is 2.0 cm×1.4 cm×1.05 cm. 1 000-grain weight of 1 400–1 500 g. PNF 25.21% and PKN 79.8% normally. Mainly cultivated in Lingchuan, Guangxi and Zhuji, Zhejiang.

图 3.119 '李子果'
① 广西桂林海洋乡40年生实生大树；
② 结果枝生长状态

Fig. 3.119 'Lizi Guo'
① 40 years old tree from seedling in Guilin, Guangxi,
② Bearing shoots

银杏核用品种——中子品种群
Zhongzi Group for Nut-producing

图 3.120 '李子果'
① 种实和种核；
② 种实和叶子；
③ 结果枝蜡叶标本

Fig. 3.120 'Lizi Guo'
① Fruits and nuts,
② Fruits and leaves,
③ Wax-leaf specimens of bearing shoots

3 '魁金'

又名'大金果'。核形系数1.34。叶缘呈大波浪状。种实倒卵形，种柄长而弯曲。核长形，上下两端似金坠，但中隐线明显，俗称'二节头'。核长×宽×厚为 2.39 cm × 1.78 cm × 1.48 cm，种壳厚0.74 mm。千粒重3 571 g，出核率28%，出仁率79%。主产于山东郯城。

'Kui Jin'

'Da Jin Guo', 'Er Jie Tou'.

RLW 1.34. Leaves margin sinuate. Fruits obovoid; stalks longer and slender. Nuts long ellipsoid, golden eardrop-shaped, with a conspicuous ridge in the middle, 2.39 cm long, 1.78 cm wide, 1.48 cm thick, sclerotesta 0.74 mm thick. 1 000-grain weight of 3 571 g. PNF 28% and PKN 79% normally. Mainly cultivated in Tancheng, Shandong.

图 3.121 '魁金'
山东郯城清泉寺林场嫁接树银杏种质基因库魁金结果枝生长状态

Fig. 3.121 'Kui Jin' Grafred tree's bearing shoots in Ginkgo resources gene bank in Tancheng, Shandong

银杏核用品种——中子品种群
Zhongzi Group for Nut-producing

图 3.122 '魁金'
①种实和种核；
②种实和叶子；
③结果枝蜡叶标本

Fig. 3.122 'Kui Jin'
① Fruits and nuts,
② Fruits and leaves,
③ Wax-leaf specimens of bearing shoots

4 '魁铃'

又名'马铃3号'。核形系数1.36。叶缘波状明显。成龄树叶长×宽为5.96 cm×8.06 cm，叶柄长5.88 cm。种实基部非正托，阔椭圆形；核肥厚，中隐线稍明显，基部两束迹间石质相连，稍歪。千粒重4 000 g。核长×宽×厚为2.65 cm×1.95 cm×1.61 cm。种壳厚0.77 mm，出核率23.7%，出仁率77.88%。主产于山东郯城。

'Kui Ling'

'Maling-3'.

RLW 1.36. Leaves margin conspicuous repand, up to 5.96 cm long, 8.06 cm wide, petiole 5.88 cm long. Fruits broad ellipsoid, oblique at the base. Nuts larger, with a slight conspicuous thin ridge in the middle, 2.65 cm long, 1.95 cm wide, 1.61 cm thick, sclerotesta 0.77 mm thick, adnate vascular bundle scars oblique at the base. 1 000-grain weight of 4 000 g. PNF 23.7% and PKN 77.88 %. Mainly cultivated in Tancheng, Shandong.

图 3.123 '魁铃'
山东郯城清泉寺林场银杏种质基因库结果枝生长状态

Fig. 3.123 **'Kui Ling'** Bearing shoots in Ginkgo resources gene bank in Tancheng, Shandong

银杏核用品种——中子品种群
Zhongzi Group for Nut-producing

图 3.124 '魁铃'
①种实和种核；
②种实和叶子；
③结果枝蜡叶标本

Fig. 3.124 'Kui Ling'
① Fruits and nuts,
② Fruits and leaves,
③ Wax-leaf specimens of bearing shoots

5 '圆底果'

又称'圆底佛手'。核形系数1.33。本品种种实先端明显平阔，呈圆底状，故名。树冠多圆锥形，树干挺直，层性明显。长枝上多扇形叶，具明显中裂。种实短卵圆形，先端圆秃，顶部平阔。顶尖凹陷呈"O"字形，珠孔迹明显。中部以下略狭缩。蒂盘圆形，表面与周缘不整，略凹陷。种柄长2.8～3.5 cm，较细弱。熟时淡黄色，被薄白粉，并可见少量油胞。种实大小为纵径2.8 cm，横径2.6 cm。种核广卵圆形，先端浑圆，具突尖，中部具不明显横脊，基部两维管束迹迹点小，相距较近，约1.5 mm，或合为一体。两侧棱线不显，仅上部可见。种核大小为1.90cm×1.43 cm×1.25 cm，单粒千核重2 450～2 680 g。出核率22%，出仁率74%。主要分布于广西灵川、浙江诸暨等。

'Yuandi Guo'

'Yuandi Foshou'.

RLW 1.33. Crown spheroidal, main trunk straight. Leaves fan-shaped on long branchlets, with conspicuous 2-lobed. Fruits short ovoid, blunt and broad with conspicuous micropylar and cusps at the apex, flattened at the tip, sunk in O-form. The base of stalk round or approximate elliptic, the surface scraggly, and the circumferentia slight concave. Stalks 2.8–3.5 cm long. Fruits light yellow and covered with thin glaicous and little hyaloid oil sac when ripe; fruits 2.8 cm long, 2.6 cm wide. Nuts broad ovoid, the apex round and with cusps, unconspicuous constriction at the midst, with two little vascular bundle scars with the average distance 1.5 mm or adnate at the base; nuts with wing-shaped raphe from the midst to the apex. Size of nut is 1.90 cm×1.43 cm×1.25 cm. 1 000-grain weight of 2 450–2 680 g. PNF 22% and PKN 74% normally. Mainly cultivated in Lingchuan, Guangxi and Zhuji of Zhejiang.

图 3.125
'圆底果'
①广西桂林海洋乡30年生实生大树；
②结果枝生长状态

Fig. 3.125 'Yuandi Guo'
① 30 years old tree from seedling in Guilin, Guangxi,
② Bearing shoots

银杏核用品种——中子品种群
Zhongzi Group for Nut-producing

图 3.126 '圆底果'
① 种实和种核；
② 种实和叶子；
③ 结果枝蜡叶标本

Fig. 3.126 'Yuandi Guo'
① Fruits and nuts,
② Fruits and leaves,
③ Wax-leaf specimens of bearing shoots

6 '桂049号'

核形系数1.30。种实长圆形，成熟时为淡黄色，表皮光滑，无油胞，有一层白粉。种实先端较圆，种柄长2.5～3.0 cm。种柄细，平直，无弯曲。种核为卵形，中部以上宽于下部，且棱线较明显，顶端圆钝，顶部凹入，并有一小尖与棱线同等高。基部两维管束迹点明显，迹点间距2.5～3.0 mm，无鱼尾凸尖，核粒较大，千粒重4 300 g。出核率26.73%，出仁率为77%。主产于广西兴安。

'Gui-049'

RLW1.30. Fruits long spherical, orange yellow and glaicous when ripe; stalks 2.5–3.0 cm long. Nuts ovoid, the apex rounded and emarginate, conspicuous raphe, vascular bundle scars separate conspicuously, range from 2.5 to 3.0 mm, without fishtail-shaped cusp.1 000-grain weight of 4 300 g. PNF 26.73% and PKN 77% normally. Mainly cultivated in Xing'an, Guangxi.

图 3.127
'桂049号'
①广西桂林海洋乡嫁接树；
②结果枝生长状态

Fig. 3.127
'Gui-049'
① Grafted tree in, Guilin, Guangxi,
② Bearing shoots

银杏核用品种——中子品种群
Zhongzi Group for Nut-producing

图 3.128 '桂 049 号'
① 种实和种核；
② 种实和叶子；
③ 结果枝蜡叶标本

Fig. 3.128 'Gui-049'
① Fruits and nuts,
② Fruits and leaves,
③ Wax-leaf specimens of bearing shoots

7 '观音皇'

又名'安陆64号'。核形系数1.40。树形松散,高15 m,胸径0.45 m。叶多扇形,叶色黄绿,叶缘缺刻较深,中裂明显。种实长圆形,熟时浅橙黄色,具少量白粉。先端圆钝,顶微凹,具小尖,珠孔迹明显,基部平。蒂盘略呈圆形,周缘不整,略凹入,种柄弯曲带钩。单粒果重13.3 g,种核长圆形似马铃,壳白饱满。先端圆钝,基部狭窄,顶具钝尖,束迹迹点小而明显,两侧棱线在1/2处消失,无背腹之分。种核千粒重3 250 g。出核率24.4%,出仁率74.1%。主产于湖北安陆。

'Guanyin Huang'

'Anlu-64'

RLW 1.40. Leaves fan-shaped, yellow green, conspicuous 2-lobed, margin repand and emarginated. Fruits long ovoid, blunt and round, with conspicuous cusps and micropylar at the apex, foveolate, flattened at the base. The base of stalk round, surface scraggly, and circumferentia slight concave. Fruits light orange yellow and covered with thin glaicous when ripe. Nuts long ovoid, white and smooth, the apex blunt and round with cusps, with adnate vascular bundle scars at the base; nut with raphe from one half of nut to the apex. 1 000-grain weight of 3 250 g. PNF 24.4% and PKN 74.1% normally. Mainly cultivated in Anlu, Hubei.

图 3.129
'观音皇'
① 湖北安陆60年生实生大树;
② 结果枝生长状态

Fig. 3.129
'Guanyin Huang'
① 60 years old tree from seedling in Anlu, Hubei,
② Bearing shoots

图 3.130 '观音皇'
① 种实和种核；
② 种实和叶子；
③ 结果枝蜡叶标本

Fig. 3.130 'Guanyin Huang'
① Fruits and nuts,
② Fruits and leaves,
③ Wax-leaf specimens of bearing shoots

8 '海洋皇'

又名'海洋王'。核形系数1.48。海洋皇原株在广西灵川海洋乡江尾村。树龄150年生，树高19 m，干高1.5 m，胸径89.7 cm，冠幅12 m×13.5 m。该品种树形开张。短枝上一般有叶片5~7枚。叶色浓绿，叶片大，长枝上的叶片中裂明显，短枝上的叶中裂不明显，缘具波状缺刻。定型叶叶长×宽为4.7~5.9 cm×6.9~8.0 cm，叶柄长4 cm。种实椭圆形，先端钝圆，顶端下凹，珠孔孔迹明显。基部平阔，蒂盘略凹陷，正托。熟时橙黄色，白粉多。种核广椭圆形，上宽下窄，腰部鼓起，核大而胖。种核先端圆，顶端不具明显小尖。基部两束迹间石质相连并突出于核之体外，形如鱼尾。两侧具棱，中上部明显。种柄长3.8~4.4 cm，上粗下细，略见弯曲，种核长×宽×厚为2.81 cm×1.90 cm×1.47 cm。千粒重3 600 g。出核率25%，出仁率77.4%。

'Haiyang Huang'

'Haiyang Wang'.
RLW 1.48. The tree grows in Haiyang, Lingchuan, Guangxi Region, 150 years old, 19 m high, 89.7 cm dbh. Crown spheroidal with 12m × 13.5 m. Short branchlets with 5-7 leaves, Leaves larger, margin repand and emarginated, conspicuous 2-lobed on long branchlets and inconspicuous 2-lobed on short branchlets, 4.7-5.9 cm long, 6.9-8.0 cm wide, petioles 4 cm long, Leave deep green. Fruits ellipsoid, blunt or rounded at the apex, excavated, with conspicuous micropylar scar at the tip, flattened at the base, the base of stalk emarginate, erect, orange yellow and glaicous when ripe. Nuts larger, broadly ellipsoid, attenuate from the apex to the base;round at the apex without cusp, adnate and protuberant vascular bundle scars at the base; fishtail-shaped, raphe conspicuous from the midst to the apex, stalks 3.8-4.4 cm long. Size of nut is 2.81 cm×1.90 cm×1.47 cm. 1 000-grain weight of 3 600 g, PNF 25% and PKN 77.4% normally.

图 3.131 '海洋皇'
①广西桂林30年生实生大树；
②结果枝生长状态

Fig. 3.131 'Haiyang Huang'
① 30 years old tree from seedling in Guilin, Guangxi;
② Bearing shoots

银杏核用品种——中子品种群
Zhongzi Group for Nut-producing

图 3.132 '海洋皇'
①种实和种核；
②种实和叶子；
③结果枝蜡叶标本

Fig. 3.132
'Haiyang Huang'
① Fruits and nuts,
② Fruits and leaves,
③ Wax-leaf specimens of bearing shoots

9 '宇香'

又名'铁富马铃3号',核形系数1.47。生长势旺盛,标准叶为扇形,叶片大而厚,叶色浓绿,中裂较浅。种实倒卵圆形,色绿黄,白粉较厚,油胞明显。种核宽卵圆形,光滑洁白,两侧棱线明显,中部以上尤显,种核最宽处具明显翼状边缘。种核千粒重3 450 g,种皮较薄,出仁率为80.2%。产于江苏邳州。

'Yuxiang'

'Tiefu Maling-3'.

RLW 1.47. Leaves fan-shaped or triangle, conspicuous 2-lobed, longer and thicker, deep color. Fruits obovoid, green yellow and cover with thick glaicous and oil sac when ripe. Nuts ovoid, symmetrical, white and smooth, with conspicuous raphe and raphe wing-shaped at the midst. 1 000-grain weight of 3 450 g. PKN 80.2% .Mainly cultivated in Pizhou, Jiangsu.

图 3.133 '宇香'(后株)
①江苏邳州嫁接树;②结果枝生长状态

Fig. 3.133 'Yuxiang'
① Grafted tree in Pizhou, Jiangsu, ② Bearing shoots

银杏核用品种——中子品种群
Zhongzi Group for Nut-producing

图 3.134 '宇香'
① 种实和种核；
② 种实和叶子；
③ 结果枝蜡叶标本

Fig. 3.134
'Yuxiang'
① Fruits and nuts,
② Fruits and leaves,
③ Wax-leaf specimens of bearing shoots

10 '马铃5号'

又名'新村5号'或'郯城5号',核形系数1.33。母树生长在山东郯城新村乡新一村,树龄150年生,树高9 m,嫁接树,树冠开心形,偏冠。叶子反卷呈喇叭状,叶色浅绿,叶长×宽为4.9 cm×6.3 cm,叶柄长4.3 cm,种实倒卵圆形或椭圆形,先端下陷,油胞小、密生,分布较均匀。种柄长3.3 cm,果长×宽×厚为2.75 cm×2.43 cm×2.39 cm,单果重11.15 g。成熟橙黄色。种核白色,椭圆形,种壳较粗糙,背腹各具3～7个小麻点,棱线上3/5明显,先端有尖,偏歪。两维管束合二为一。种核中隐线明显。核长×宽×厚为2.44 cm×1.83 cm×1.31 cm。千粒重2 440 g。出核率24%,出仁率78%。

'Maling-5'

'Xincun-5', 'Tancheng-5'.

RLW 1.33. The tree grows in Xincun, Tancheng, Shandong, grafting tree, 150 years old, 9 m height with open center-shaped and lopsided crown. Leaves inrolled and trumpet-shaped, 4.9 cm long, 6.3 cm wide, petioles 4.3 cm long, light green. Fruits obovoid or ellipsoid, emarginate at the apex, with smaller and density ocellus, 2.75 cm long, 2.43 cm wide, 2.39 cm thick; stalks 3.3 cm long. Nuts ellipsoid, foveolates (3–7) and muriculate, raphe conspicuous, with cusp at the apex, slight oblique, adnate vascular bundle scars at the base, a conspicuous ridge at the middle, Size of nut is 2.44 cm×1.83 cm×1.3l cm. 1 000-grain weight of 2 440 g. PNF 24% and PKN 78% normally.

图 3.135
'马铃5号'
山东郯城清泉寺林场银杏种质基因库结果枝生长状态

Fig. 3.135
'Maling-5'
Bearing shoots in Ginkgo resources gene bank in Tancheng, Shandong

银杏核用品种——中子品种群
Zhongzi Group for Nut-producing

图 3.136
'马铃5号'
① 种实和种核；
② 种实和叶子；
③ 结果枝蜡叶标本

Fig. 3.136 '**Maling-5**'
① Fruits and nuts,
② Fruits and leaves,
③ Wax-leaf specimens of bearing shoots

11 '梅核'

又称'大梅核'、'小梅核'。本品种之种核似"梅子"之核，故名梅核。核形系数1.53。本品种在全国各地广泛分布。树冠圆锥形，层性明显，生长势强，长枝多扇形叶，少数为三角形叶和截形叶，叶片均具明显中裂，一般裂深达2.0 cm以上，最深可达3.0 cm以上，叶缘具波状缺刻。种实短椭圆形，先端渐尖，具小尖头，珠孔迹明显，多不凹入。基部平，稍见收缩，蒂盘长椭圆形，较小，周缘不整，凹入于种皮中。种柄较短，约2.5~3.1 cm，下部宽扁，略有弯曲。熟时杏黄色，被薄白粉，表皮有皱褶，并具梭状油胞。大梅核，种实大小约纵径2.87 cm，横径2.2 cm，单粒种实重可达11.25 g。小者称小梅核，种实大小约纵径2.64 cm，横径2.16 cm。种核椭圆状纺锤形，上下两端基本一致，洁白光亮，腰部浑圆丰满，少数有腹背之分。先端稍宽圆，具微尖。基部光洁度差，稍现粗糙，两束迹迹点显明，相距较近，不足2 mm，有时中间相连呈短横线状，个别的聚为一点。两侧棱线明显，中上部成窄其状边缘。种核大小为2.35 cm×1.53 cm×1.29 cm。千粒重1 825 g，出核率均在25%左右，出仁率78.0%。

'Meihe'

'Da Meihe', 'Xiao Meihe'.

Nuts plum corn-shaped. RLW1.53. Crown spheroidal, main trunk straight. Leaves fan-shaped on long branchlets, few triangle and truncate, leaves with conspicuous 2-lobed, margin repand and emarginated. Fruits short ellipsoid, with cusps and conspicuous micropylar at the apex, flattened at the base. The base of stalk long ellipse, surface scraggly, and circumferentia slight concave. Stalks average 2.5–3.1 cm long, slight oblique. Fruits apricot and covered with many oil sac and thin glaicous when ripe. Da Meihe 2.87 cm long, 2.2 cm wide, Xiao Meihe 2.64 cm long, 2.16 cm wide. Nuts ellipsoid, symmetrical, white, with cusps at the apex, with two conspicuous vascular bundle scars at the base with the average distance 2.0 mm. Nuts with conspicuous raphe. Size of nut is 2.35 cm×1.53 cm×1.29 cm. 1 000-grain weight of 1 825 g. PNF 25% and PKN 78.0% normally.

图 3.137 '梅核'
① 江苏邳州陈楼银杏种质基因库 15 年生嫁接树；
② 结果枝生长状态

Fig. 3.137 'Meihe'
① 15 years old grafted tree in Ginkgo resources gene bank in Pizhou, Jiangsu,
② Bearing shoots

银杏核用品种——中子品种群
Zhongzi Group for Nut-producing

图3.138 '梅核'
①种实和种核；
②种实和叶子；
③结果枝蜡叶标本

Fig. 3.138 'Meihe'
① Fruits and nuts,
② Fruits and leaves,
③ Wax-leaf specimens of bearing shoots

12 '棉花果'

本品种种实形状近似棉铃，故名。核形系数1.39。树冠圆锥形。主干挺直，层性明显。长枝节间长达4.0 cm以上。长枝上多扇形叶，少数为三角形，叶片中裂明显。种实椭圆形，多双果，熟时橙黄色，白粉少，无油胞。先端狭长，顶尖凹入，基部稍平，蒂盘多长圆形，周缘不整，稍见凹陷。种实柄长约3.6～4.3 cm，较粗壮，略弯曲。种实大小为纵径2.6 cm，横径2.4 cm，单粒种实平均重9.4 g，每千克粒数106粒。种实整齐度相差悬殊，有的种实单粒重仅6.67 g，每千克粒数达150粒。种核椭圆形，上下基本对称。先端尖长，顶具小尖。基部两维管束迹迹点明显，相距甚近，间距仅约1.5 mm，有时合为一体。两侧棱线明显，自上至下均有。微有背腹之分。种核大小为2.15 cm×1.55 cm×1.26 cm，种核千粒重2 100 g。出核率23.18%，出仁率75.56%。本品种主要分布于广西的灵川、兴安、全州。

'Mianhua Guo'

Fruits cotton boll-shaped. RLW 1.39. Crown spheroidal, main trunk straight. Leaves fan-shaped on long branchlets, few triangle, leaves with conspicuous 2-lobed. Fruits ellipsoid, long and narrow at the apex, flattened at the base. The base of stalk oblong, surface scraggly and circumferentia slight concave. Stalks average 3.6–4.3cm long, slight oblique. Fruits orange yellow and covered with thin glaicous when ripe, fruits 2.6 cm long, 2.4 cm wide. Nuts ellipsoid, symmetrical, white, sharp and narrow, with cusps at the apex, with two conspicuous vascular bundle scars at the base with the average distance 1.5 mm or adnate. Nuts with conspicuous raphe. Size of nut is 2.15 cm×1.55 cm×1.26 cm.1 000-grain weight of 2 100 g. PNF 23.18% and PKN 75.56%. Mainly cultivated in Lingchuan, Xing'an, Quanzhou of Guangxi.

图 3.139 '棉花果'
①广西桂林海洋乡30年生实生大树；
②结果枝生长状态

Fig. 3.139 'Mianhua Guo'
① 30 years old tree from seadling in Guilin, Guangxi,
② Bearing shoots

银杏核用品种——中子品种群
Zhongzi Group for Nut-producing

图 3.140 '棉花果'
① 种实和种核；
② 种实和叶子；
③ 结果枝蜡叶标本

Fig. 3.140
'Mianhua Guo'
① Fruits and nuts,
② Fruits and leaves,
③ Wax-leaf specimens of bearing shoots

13 '早实梅核'

又名'23号大梅核'。核形系数1.32。熟时外种皮呈暗褐色。该品种种核无麻点，光滑，中线稍明显，背腹均圆。核长×宽×厚为2.45 cm×1.85 cm×1.58 cm，千粒重3 400 g，出核率25%，出仁率81%。主要产于湖北安陆。

'Zaoshi Meihe'

'Da Meihe-23'.

RLW 1.32. Fruits dark brown when ripe. Nuts spherical, glazed, symmetrical, a conspicuous ridge at the middle, 2.45 cm long, 1.85 cm wide, 1.58 cm thick. 1 000-grain weight of 3 400 g. PNF 25% and PKN 81% normally. Mainly cultivated in Anlu, Hubei.

图 3.141
'早实梅核'
湖北安陆60年生实生大树；结果枝生长状态

Fig. 3.141
'Zaoshi Meihe' Bearing shoots of 60 years old tree from seedling in Anlu, Hubei

银杏核用品种——中子品种群
Zhongzi Group for Nut-producing

图 3.142
'早实梅核'
① 种实和种核；
② 种实和叶子；
③ 结果枝蜡叶标本

Fig. 3.142
'Zaoshi Meihe'
① Fruits and nuts,
② Fruits and leaves,
③ Wax-leaf specimens of bearing shoots

14 '安陆1号'

又名'安陆大白果'。核形系数1.30。树势强健,层性明显。多为扇形叶,少数为三角形叶。叶色较深,叶缘缺刻较浅,中裂不明显。种实近圆球形,熟时橙黄色,具白粉。先端具小尖或微凹入,基部蒂盘小,表面凸凹不平,周边不整,稍下陷。种柄较短,稍弯曲,粗壮。种实平均单粒重13.2 g,每千克粒数76。种核长圆形且饱满,壳乳白色,光滑,先端圆秃,具小尖,维管束迹迹点宽平,两侧棱线(有的有翼)至基部4/5消失,背腹相等。种核千粒重3 700 g。出核率28%,出仁率78%,产于湖北安陆。

'Anlu-1'

'Anlu Da Baiguo'.

RLW 1.30. Leaves fan-shaped or triangle, inconspicuous 2-lobed, margin repand and emarginated, deep color. Fruits ovoid, with conspicuous cusps at the apex, foveolate. The base of stalk surface scraggly and circumferentia slight concave. Fruits orange yellow and covered with thicker glaicous when ripe. Nuts long ovoid, white and smooth, the apex blunt and round with cusps and adnate vascular bundle scars at the base, nuts with raphe from one fifth of nut to the base. 1 000-grain weight of 3 700 g. PNF 28% and PKN 78% normally. The tree grows in Wangyizhen, Anlu, Hubei.

图 3.143
'安陆1号'
结果枝生长状态

Fig. 3.143
'Anlu-1'
Bearing shoots

银杏核用品种——中子品种群
Zhongzi Group for Nut-producing

图 3.144
'安陆 1 号'
① 种实和种核；
② 种实和叶子；
③ 结果枝蜡叶标本

Fig. 3.144
'Anlu-1'
① Fruits and nuts,
② Fruits and leaves,
③ Wax-leaf specimens of bearing shoots

135

3.1.3.3 中子品种群优良无性系

分别对中子品种群中 14 个优良无性系进行简要介绍，见表 3.3。

3.1.3.3 Clones of Zhongzi Group

Table 3.3 illustrated 14 clones of Zhongzi Group in brief.

表 3.3 中子品种群优良无性系

Table 3.3 A list of Some Ginkgo Clones of Zhongzi Group

编号 Number	名称 Name	核形系数 RLW	产地 Distribution
1	'郯马 1 号' 'Tan Ma-1'	1.33	山东郯城新村 Xincun, Tancheng, Shandong
2	'正安 5 号' 'Zheng'an-5'	1.34	贵州正安 Zheng'an, Guizhou
3	'郯城 207 号' 'Tancheng-207'	1.35	山东郯城 Tancheng, Shandong
4	'新村 16 号' 'Xincun-16'	1.37	山东郯城新村 Xincun, Tancheng, Shandong
5	'铁马 1 号' 'Tie Ma-1'	1.38	江苏邳州铁富 Tiefu, Pizhou, Jiangsu
6	'安陆 A14 号' 'Anlu-A14'	1.409	湖北安陆 Anlu, Hubei
7	'郯城 9 号' 'Tancheng-9'	1.41	山东郯城新村 Xincun, Tancheng, Shandong
8	'郯城 322 号' 'Tancheng-322'	1.41	山东郯城 Tancheng, Shandong
9	'正安 3 号' 'Zheng'an-3'	1.41	贵州正安 Zheng'an, Guizhou
10	'新村 18 号' 'Xincun-18'	1.41	山东郯城新村 Xincun, Tancheng, Shandong
11	'郯城 16 号' 'Tancheng-16'	1.42	山东郯城新村 Xincun, Tancheng, Shandong
12	'安陆 A3-1 号' 'Anlu-A3-1'	1.42	湖北安陆 Anlu, Hubei
13	'桂林 8 号' 'Guilin-8'	1.42	广西桂林林科所 Guilin Forestry Academy, Guangxi
14	'港上 309 号' 'Gangshang-309'	1.44	山东郯城港上 Gangshang, Tancheng, Shandong

银杏核用品种——中子品种群
Zhongzi Group for Nut-producing

'郯马1号' 'Tan Ma-1'

图 3.145
'郯马1号'
结果枝生长状态

Fig. 3.145
'Tan Ma-1'
Bearing shoots

图 3.146
'郯马1号'
①种实和种核；
②种实和叶子；
③结果枝蜡叶标本

Fig. 3.146
'Tan Ma-1'
① Fruits and nuts,
② Fruits and leaves,
③ Wax-leaf specimens of bearing shoots

'正安5号'　'Zheng'an-5'

图 3.147
'正安5号'
结果枝生长状态

Fig. 3.147
'Zheng'an-5'
Bearing shoots

图 3.148
'正安5号'
① 结果枝蜡叶标本；
② 种实和种核；
③ 种实和叶子

Fig. 3.148
'Zheng'an-5'
① Wax-leaf specimens of bearing shoots,
② Fruits and nuts,
③ Fruits and leaves

'郯城 207 号' 'Tancheng-207'

图 3.149
'郯城 207 号'
结果枝生长状态

Fig. 3.149
'Tancheng-207'
Bearing shoots

图 3.150
'郯城 207 号'
①种实和种核；
②种实和叶子；
③结果枝蜡叶标本

Fig. 3.150
'Tancheng-207'
① Fruits and nuts,
② Fruits and leaves,
③ Wax-leaf specimens of bearing shoots

'新村16号' 'Xincun-16'

图 3.151
'新村 16 号'
结果枝生长状态

Fig. 3.151
'Xincun-16'
Bearing shoots

图 3.152
'新村 16 号'
①结果枝蜡叶标本；
②种实和种核；
③种实和叶子

Fig. 3.152
'Xincun-16'
① Wax-leaf specimens of bearing shoots,
② Fruits and nuts,
③ Fruits and leaves

'铁马1号' 'Tie Ma-1'

图 3.153
'铁马1号'
结果枝生长状态

Fig. 3.153
'Tie Ma-1'
Bearing shoots

图 3.154
'铁马1号'
① 种实和种核；
② 种实和叶子；
③ 结果枝蜡叶标本

Fig. 3.154
'Tie Ma-1'
① Fruits and nuts,
② Fruits and leaves,
③ Wax-leaf specimens of bearing shoots

'安陆A14号' 'Anlu-A14'

图 3.155
'安陆 A14 号'
结果枝生长状态

Fig. 3.155
'Anli-A14'
Bearing shoots

图 3.156
'安陆 A14 号'
①结果枝蜡叶标本；
②种实和种核；
③种实和叶子

Fig. 3.156
'Anli-A14'
① Wax-leaf specimens of bearing shoots,
② Fruits and nuts,
③ Fruits and leaves

'郯城9号' 'Tancheng-9'

图 3.157
'郯城9号'
结果枝生长状态

Fig. 3.157
'Tancheng-9'
Bearing shoots

图 3.158
'郯城9号'
①种实和种核；
②种实和叶子；
③结果枝蜡叶标本

Fig. 3.158
'Tancheng-9'
① Fruits and nuts,
② Fruits and leaves,
③ Wax-leaf specimens of bearing shoots

'郯城322号' 'Tancheng-322'

图 3.159
'郯城 322 号'
结果枝生长状态

Fig. 3.159
'Tancheng-322'
Bearing shoots

图 3.160
'郯城 322 号'
① 结果枝蜡叶标本；
② 种实和种核；
③ 种实和叶子

Fig. 3.160
'Tancheng-322'
① Wax-leaf specimens of bearing shoots,
② Fruits and nuts,
③ Fruits and leaves

144

'正安3号' 'Zheng'an-3'

图 3.161
'正安 3 号'
结果枝生长状态

Fig. 3.161
'Zheng'an-3'
Bearing shoots

图 3.162
'正安 3 号'
① 种实和种核；
② 种实和叶子；
③ 结果枝蜡叶标本

Fig. 3.162
'Zheng'an-3'
① Fruits and nuts,
② Fruits and leaves,
③ Wax-leaf specimens of bearing shoots

'新村18号' 'Xincun-18'

图 3.163
'新村 18 号'
结果枝生长状态

Fig. 3.163
'Xincun-18'
Bearing shoots

图 3.164
'新村 18 号'
①结果枝蜡叶标本；
②种实和种核；
③种实和叶子

Fig. 3.164
'Xincun-18'
① Wax-leaf specimens of bearing shoots,
② Fruits and nuts,
③ Fruits and leaves

'郯城16号' 'Tancheng-16'

图 3.165
'郯城 16 号'
结果枝生长状态

Fig. 3.165
'Tancheng-16'
Bearing shoots

图 3.166
'郯城 16 号'
① 种实和种核；
② 种实和叶子；
③ 结果枝蜡叶标本

Fig. 3.166
'Tancheng-16'
① Fruits and nuts,
② Fruits and leaves,
③ Wax-leaf specimens of bearing shoots

第3章 银杏品种检索表及品种图谱
Chapter 3　Key to Cultivars of *Ginkgo biloba* L. and Illustration of Ginkgo Cultivars

'安陆 A3-1号'　'Anlu-A3-1'

图 3.167
'安陆 A3-1 号'
结果枝生长状态

Fig. 3.167
'Anlu-A3-1'
Bearing shoots

图 3.168
'安陆 A3-1 号'
①结果枝蜡叶标本；
②种实和种核；
③种实和叶子

Fig. 3.168
'Anlu-A3-1'
① Wax-leaf specimens of bearing shoots,
② Fruits and nuts,
③ Fruits and leaves

银杏核用品种——中子品种群
Zhongzi Group for Nut-producing

'桂林 8 号' 'Guilin-8'

图 3.169
'桂林 8 号'
结果枝生长状态

Fig. 3.169
'Guilin-8'
Bearing shoots

图 3.170
'桂林 8 号'
①种实和种核；
②种实和叶子；
③结果枝蜡叶标本

Fig. 3.170
'Guilin-8'
① Fruits and nuts,
② Fruits and leaves,
③ Wax-leaf specimens of bearing shoots

149

'港上309号' 'Gangshang-309'

图 3.171
'港上 309 号'
结果枝生长状态

Fig. 3.171
'Gangshang-309'
Bearing shoots

图 3.172
'港上 309 号'
① 结果枝蜡叶标本；
② 种实和种核；
③ 种实和叶子

Fig. 3.172
'Gangshang-309'
① Wax-leaf specimens of bearing shoots,
② Fruits and nuts,
③ Fruits and leaves

3.1.4 圆子品种群

将银杏种核长与宽的比值小于1.30（±0.05），种核纵横轴的交叉点位于纵线的中心的银杏品种划分为圆子品种群，共有11个优良品种及6个优良无性系。本节重点介绍这11个优良品种特性，另有6个优良无性系只以表格和照片的形式做简要介绍。

3.1.4 Yuanzi Group

There are 11 cultivars and 6 clones of Yuanzi Group, according to the ratio of length to width less than 1.30 (±0.05), the intersection of ordinate axis and horizontal axis of nut site one half of ordinate axis.

浙江长兴古银杏群落
The Ancient Ginkgo Trees in Changxing, Zhejiang

3.1.4.1 圆子品种群分类检索表

圆子品种群检索表

1.种核两侧棱不具翼状边缘
    2.种核基部两维管束迹呈二点状或合二为一
        3.种核侧棱明显..1.'团峰'
        3.种核侧棱4/5明显...2.'郯丰'
    2.种核基部两维管束迹点明显，间距3mm左右
        4.种核先端较尖，顶点与棱线相连，小尖明显
            5.种实蒂盘略凹陷..3.'桂047号'
            5.种实蒂盘平...4.'桂048号'
        4.种核先端圆钝，中间凹下不具小尖，呈鱼嘴状，珠孔迹可见但不凸出............
            ....5.'葡萄果'
1.种核两侧棱具翼状边缘
    6.种实先端具小尖或微微凹入
        7.种核基部维管束迹迹点较小
            8.种核基部光滑，不具纵沟
                9.种实和种核上下两半不均等，基部稍宽
                    10.种核基部两维管束迹迹点间距较小，约1.0~1.5mm.................6.'龙眼'
                    10.种核基部两维管束迹迹点间距特宽，平均宽5.25mm...................7.'大圆子'
                9.种实和种核上下两半均似相等...8.'圆铃'
            8.种核基部四周具纵沟，形成的对角线或十字线纵沟明显深长...............9.'小圆子'
        7.种核基部维管束迹迹点大而明显，常一高一低，迹点间距较大..............10.'桐子果'
    6.种实先端极尖..11.'皱皮果'

3.1.4.1 Key to Yuanzi Group

Key to Yuanzi Group of *Ginkgo biloba* L.

1. The raphe not wing-shaped
    2. Nuts with two spotted or adnate vascular bundle scars at the base
        3. Nuts with conspicuous raphe .... 1. 'Tuan Feng'
        3. Four fifth of raphe conspicuous .... 2. 'Tan Feng'
    2. Nuts with conspicuous vascular bundle scars at the base with distance 3.0 mm
        4. Nuts with conspicuous cusps at the apex, the tip of nuts adnate with raphe
            5. The base of stalk emarginate .... 3. 'Gui-047'
            5. The base of stalk flattened .... 4. 'Gui-048'
        4. Nuts blunt and round, foveolate but without cusp, fish mouth-shaped, with unconspicuous micropylar .... 5. 'Putao Guo'
1. Nuts with conspicuous wing-shaped raphe
    6. Fruits with cusps or slight foveolate at the apex
        7. Nuts with small vascular bundle scars at the base
            8. Nuts without stria at the base
                9. Fruits and nuts asymmetrical, the base wider than the apex
                    10. Nuts with vascular bundle scars at the base with distance 1.0~1.5mm ..... 6. 'Longyan'
                    10. Nuts with vascular bundle scars at the base with distance 5.25mm ...... 7. 'Da Yuanzi'
                9. Fruits and nuts symmetrical .... 8. 'Yuanling'
            8. Nuts with stria at the base, the diagonal or cross-line stria conspicuous .... 9. 'Xiao Yuanzi'
        7. Nuts with conspicuous and large vascular bundle scars at the base with long distance .... 10. 'Tongzi Guo'
    6. Fruits with sharp apex .... 11. 'Zhoupi Guo'

3.1.4.2 圆子品种群优良品种
分别对圆子品种群 11 个优良品种进行详细介绍。

3.1.4.2 Key to Yuanzi Group
Illustrated 11 cultivars of Yuanzi Group in detail.

1 '团峰'

又称'大龙眼'或'圆铃6号'。核形系数1.15。叶长×宽为5.08 cm×8.02 cm，叶柄长5.88 cm，叶缘波状或二裂。种实圆形，正托。核肥厚圆形、规整，侧棱明显，基部两维管束迹呈二点状或合二为一。千粒重3 040 g，种壳厚0.52 mm。核长×宽×厚为2.05 cm×1.78 cm×1.42 cm。出核率24%，出仁率82%。主产于山东郯城。

'Tuan Feng'

'Da Longyan', 'Yuanling-6'.

RLW 1.15. Leaves 2-lobed, margin repand, 5.08 cm long, 8.02 cm wide, petiole 5.88 cm long. Fruits spheroidal, erect at the base. Nuts spheroidal, symmetrical and regular, raphe conspicuous, vascular bundle scars adnate or two punctual. Sclerotesta 0.52 mm thick. Size of nut is 2.05 cm×1.78 cm×1.42 cm. 1 000-grain weight of 3 040 g. PNF 24% and PKN 82% normally. Mainly cultivated in Tancheng, Shandong.

图 3.173 '团峰'
①山东郯城清泉寺嫁接树；
②结果枝生长状态

Fig. 3.173 'Tuan Feng'
① Grafted tree in Qingquan Temple in Tancheng, Shandong, ② Bearing shoots

银杏核用品种——圆子品种群
Yuanzi Group for Nut-producing

图 3.174 '团峰'
① 种实和种核；
② 种实和叶子；
③ 结果枝蜡叶标本

Fig. 3.174
'Tuan Feng'
① Fruits and nuts,
② Fruits and leaves,
③ Wax-leaf specimens of bearing shoots

2 '郯丰'

又名'郯城107号'。核形系数1.21。母树在重坊镇埔里村,嫁接树,树龄35年生,开心形树冠。干高2.2 m,树高9 m,胸径24 cm,生长健壮。叶缘呈小波浪形,微卷。种实呈阔椭圆形,顶端平广微凹,基部平广,形态成熟时外表杏黄色,果粉中等,种柄正托,果蒂椭圆形,油胞红褐色,较少。果长×宽×厚为2.71 cm×2.51 cm×2.50 cm,种柄长2.9 cm。种核呈阔椭圆形,顶端具尖,基部两维管束合二为一,侧棱上4/5明显,种壳呈象牙白。种壳厚0.56 mm,千粒重2 580 g。出核率25%,出仁率79%。主产于山东郯城。

'Tan Feng'

'Tancheng-107'.

RLW 1.21. The tree grows in Chongfang, grafting tree, 35 years old, up to 9 m height, 24 cm dbh, with open center-shape crown. Leaves margin slightly inrolled and repand. Fruits broad ellipsoid, flattened and slight emarginate at the apex, broaden and flattened at the base, with a few mahogany ocellus, apricot when ripe; stalk erect, 2.9 cm long, the base of stalk elliptic. Nuts broadly ellipsoid, with cusp at the apex, raphe conspicuous, adnate vascular bundle scars at the base, sclerotesta 0.56 mm. Size of nut is 2.71 cm×2.51 cm×2.50 cm. 1 000-grain weight of 2 580 g. PNF 25% and PKN 79% normally. Mainly cultivated in Tancheng, Shandong.

图 3.175 '郯丰'
山东郯城清泉寺林场银杏种质基因库魁金结果枝生长状态

Fig. 3.175 'Tan Feng' Bearing shoots in Ginkgo resources gene bank in Tancheng, Shandong

银杏核用品种——圆子品种群
Yuanzi Group for Nut-producing

图 3.176 '郯丰'
① 种实和种核；
② 种实和叶子；
③ 结果枝蜡叶标本

Fig. 3.176
'Tan Feng'
① Fruits and nuts,
② Fruits and leaves,
③ Wax-leaf specimens of bearing shoots

3 '桂047号'

核形系数1.16。种实为圆形,先端微凹,呈"一"形,有小尖,蒂盘略凹陷,种柄长3.4~3.5 cm,上粗下细,略弯曲。成熟时为橙黄色,有一层白粉,油胞明显。种核核粒较大,为椭圆形,先端较尖,顶点小尖明显,并与两侧棱线连接,棱线偏中下部往上。种核饱满,基部两维管束迹点明显,一高一低,两点间距3~4 mm,无鱼尾状凸尖。千粒重3 500 g,出核率26%,出仁率76%。核仁为白色。主产于广西兴安。

'Gui-047'

RLW 1.16. Fruits spherical, emarginated with cusp at the apex. The base of stalk emarginated; stalks 3.4–3.5 cm long, oblique and attenuate, orange yellow with glaicous and conspicuous ocellus when ripe. Nuts larger, ellipsoid, apiculate at the apex with conspicuous cusps, adnate with raphe, conspicuous vscular bundle scars at the base with 3–4 mm distance and not fishtail-shaped cusp. Kernels white. 1 000-grain weight of 3 500 g. PNF 26% and PKN 76% normally. Mainly cultivated in Xing'an, Guangxi.

图 3.177
'桂047号'
①广西桂林海洋乡10年生嫁接树;
②结果枝生长状态

Fig. 3.177
'Gui-047'
① 10 years old grafted tree in Guilin, Guangxi,
② Bearing shoots

银杏核用品种——圆子品种群
Yuanzi Group for Nut-producing

图 3.178
'桂 047 号'
① 种实和种核；
② 种实和叶子；
③ 结果枝蜡叶标本

Fig. 3.178
'Gui-047'
① Fruits and nuts,
② Fruits and leaves,
③ Wax-leaf specimens of bearing shoots

4 '桂048号'

核形系数1.18。种实为圆形，先端凹，有小尖，成熟时为橙黄色，有一层白粉，油胞明显。蒂盘平，种柄长3.2～3.4 cm，略弯曲。种实纵径3.0 cm，横径2.8 cm，平均单粒重8.9 g，每千克112粒。种核为椭圆形，先端较尖，顶点与棱线相连，小尖明显，棱线自下往上明显，种核饱满，基部两维管束迹点明显，略见鱼尾状，两束迹间距3 mm左右，核粒较大，种实千粒重3 300 g，出核率26%，出仁率为77.5%。主产于广西兴安。

'Gui-048'

RLW1.18. Fruits spheroidal, the apex emarginate, with a cusps, orange yellow and glaicous, covered conspicuously ocellus; the base of stalk flattened, stalks 3.2–3.4 cm long. Nuts ellipsoid, with conspicuous cusps at the apex, vascular bundle scars separate conspicuously and fishtail-shaped. 1 000-grain weight of 3 300 g. PNF 26% and PKN 77.5%. Mainly cultivated in Xing'an, Guangxi.

图 3.179
'桂048号'
① 广西桂林海洋乡10年生嫁接树；
② 结果枝生长状态

Fig. 3.179
'Gui-048'
① 10 years old grafted tree in Guilin, Guangxi,
② Bearing shoots

银杏核用品种——圆子品种群
Yuanzi Group for Nut-producing

图 3.180
'桂 048 号'
① 种实和种核；
② 种实和叶子；
③ 结果枝蜡叶标本

Fig. 3.180
'Gui-048'
① Fruits and nuts,
② Fruits and leaves,
③ Wax-leaf specimens of bearing shoots

5 '葡萄果'

本品种的结实能力强，挂果成串，挤满枝条，形似葡萄，故名。核形系数1.16。树冠圆头形或圆锥形，无中心主干，侧枝分布均匀。长枝上多扇形叶，少数为三角形叶，中裂明显，叶色深，叶片大。种实圆球形，先端平圆，顶凹入呈"一"字形，珠孔迹明显。基部圆形，蒂盘呈不规则长圆形，周缘不整，稍凹陷。种实柄长3.6 cm，可达4.5 cm，略弯曲。种实大小为纵径2.77 cm，横径2.79 cm，熟时淡黄色，被厚白粉，有透明梭状油胞。种核圆形，先端圆钝，中间凹下不具小尖，呈鱼嘴状，珠孔迹可见但不凸出。基部平，稍呈狭长，两维管束迹迹点明显，一高一低，间距大，3.2~4.3 mm。两侧棱线明显，自上而下均见。有背腹之分但不明显。种核大小为2.15 cm×1.86 cm×1.43 cm，种实千粒重2 510 g。出核率21.7%，出仁率80.1%。主要分布于广西桂林。

'Putao Guo'

Fruits grape-shaped. RLW 1.16. Crown spheroidal or spherical. Leaves fan-shaped on long branchlets and short branchlets leaves fan-shaped or triangle, with conspicuous 2-lobed, deep colour. Fruits ovoid, flattened and round, foveolate, with conspicuous micropylar at the apex, round at the base. The base of stalk approximate oblong and the circumferentia slight concave. Stalks 3.6 cm long and slight oblique; Fruits average 2.77 cm long, 2.79 cm wide. Fruits light yellow and covered with thicker glaicous and hyaloid oil sac when ripe. Nuts ovoid, blunt and round, foveolate but without cusp, fish mouth-shaped, with unconspicuous micropylar. Nuts base flattened, vascular bundle scars at the base with distance about 3.2–4.3 mm, nuts with conspicuous raphe.1 000-grain average weight of 2 510 g. Size of nut is 2.15 cm×1.86 cm×1.43 cm. PNF 21.7% and PKN 80.1% normally. Mainly cultivated in Guilin, Guangxi.

图 3.181
'葡萄果'
①广西桂林海洋乡30年生实生大树；
②结果枝生长状态

Fig. 3.181
'Putao Guo'
① 30-year-old tree from seedling in Guilin, Guangxi, ② Bearing shoots

银杏核用品种——圆子品种群
Yuanzi Group for Nut-producing

图 3.182
'葡萄果'
① 种实和种核；
② 种实和叶子；
③ 结果枝蜡叶标本

Fig. 3.182
'Putao Guo'
① Fruits and nuts,
② Fruits and leaves,
③ Wax-leaf specimens of bearing shoots

6 '龙眼'

近似无患子科之龙眼，故名。核形系数1.15。叶片多扇形，少数三角形，具中裂。叶色深，叶片厚，叶脉粗壮。种实卵圆形，先端圆钝，顶部呈"O"字形凹入，珠孔孔迹明显。基部蒂盘长椭圆形，表面高低不平，周缘波状，稍见凹入。种柄长短不一，长1.0～4.5 cm。熟时橙黄色，具厚白粉。多双果。种核卵圆形，略扁，中间鼓起，丰满状。先端钝圆，具不明显之小尖。基部维管二束迹迹点较小，但明显突出，两迹点间距约1.0～1.5 mm。两侧棱线明显且可见宽翼状边缘。种核大小为1.5 cm×1.35 cm×1.1 cm，千粒重仅1 500 g，出核率仅约17%。

'Longyan'

Nuts longyan-shaped. RLW 1.15. Leaves fan-shaped or triangle, conspicuous 2-lobed, longer and thicker, deep color. Fruits ovoid, blunt and broad with conspicuous micropylar and cusps at the apex, blunt and round at the tip, emarginated and O form. Stalks 1.0–4.5 cm long, fruits orange yellow and cover with thick glaicous when ripe. Nuts ovoid, symmetrical, blunt and broad with unconspicuous cusp at the apex, with two little vascular bundle scars at the base with distance 1.0–1.5 mm. Size of nut is 1.5 cm×1.35 cm×1.1 cm. 1 000-grain weight of 1 500 g. PNF 17%.

图 3.183 '龙眼'
① 江苏邳州嫁接树；
② 结果枝生长状态

Fig. 3.183 **'Longyan'**
① Grafted tree in Pizhou, Jiangsu,
② Bearing shoots

银杏核用品种——圆子品种群
Yuanzi Group for Nut-producing

图 3.184 '龙眼'
①种实和种核；
②种实和叶子；
③结果枝蜡叶标本

Fig. 3.184
'Longyan'
① Fruits and nuts,
② Fruits and leaves,
③ Wax-leaf specimens of bearing shoots

7 '大圆子'

又称'大圆珠'、'大圆头'。核形系数1.12。树冠直立，大枝平展，老枝下垂，与主干夹角大。叶多三角状扇形，叶基夹角约70°，叶面稍向上纵卷，具浅中裂或不明显。缘具波状浅缺刻，大小约5.3 cm×4.3 cm。种实圆球形，先端圆钝，基部平广，蒂盘圆形或椭圆形，平或稍凹，具细浅皱纹。种柄短，长1.87～3.44 cm，上细下粗，近蒂盘处粗约0.29 cm。种实大小约纵径3.03 cm，横径3.17 cm，单粒种实平均重约17.3 g，每千克粒数58粒。熟时淡黄色，稍带红晕，被薄白粉。种核圆形，腹背面不显。先端圆钝，具不明显之细小顶尖，基部稍宽，近束迹处明显变狭，迹点细小，间距特宽，平均宽5.25 mm，与'龙眼'种核显然不同。两侧棱线明显，呈翼状，但基部其宽而顶部翼狭。种核大小为2.47 cm×2.19 cm×1.74 cm，种核千粒重4 080 g。出核率23.3%，出仁率77.4%。

'Da Yuanzi'

'Da Yuanzhu', 'Da Yuantou'.

RLW 1.12. Crown erect. Leaves triangle or fan-shaped, with shallow 2-lobed, margin emarginated, 5.3 cm long, 4.3 cm wide. Fruits ovoid, blunt and round at the apex, flattened at the base. The base of stalk round or ellipse, the circumferentia slight concave, with shallow stria. Stalks short, 1.87–3.44 cm long. Fruits 3.03 cm long, 3.17 cm wide, fruits light yellow and covered with thin glaicous when ripe. Nuts spheroidal, asymmetrical, blunt and round, with unconspicuous cusps at the apex, flattened at the base, two little vascular bundle scars with the average distance 5.25 mm, nuts with wing-shaped raphe. Size of nut is 2.47 cm×2.19 cm×1.74 cm. 1 000-grain weight of 4 080 g. PNF 23.3% and PKN 77.4% normally.

图 3.185
'大圆子'
广西桂林海洋乡大圆子结果枝生长状态

Fig. 3.185
'Da Yuanzi'
Bearing shoots in Guilin, Guangxi

银杏核用品种——圆子品种群
Yuanzi Group for Nut-producing

图 3.186
'大圆子'
① 种实和种核；
② 种实和叶子；
③ 结果枝蜡叶标本

Fig. 3.186
'Da Yuanzi'
① Fruits and nuts,
② Fruits and leaves,
③ Wax-leaf specimens of bearing shoots

8 '圆铃'

核形系数1.20。树冠呈塔形,有明显的中心主干。长枝梢端多三角形叶,中下部则多扇形及截形叶,叶片具明显中裂,中上部叶片中裂尤深。种实圆球形或近圆球形,先端具小尖或微微凹入。种实及种核上下两半均近相等。基部蒂盘小,表面凹凸不平,周边不整且稍下陷。种柄长约3.1 cm。种实大小为纵径3.1 cm,横径2.96 cm(小者2.4 cm×2.34 cm),单粒种实重11.8~15.7 g,每千克粒数64~85粒,熟时青黄色,被薄白粉。种核卵圆形,丰满。先端突出具小尖。基部束迹迹点明显,但迹点甚小,迹点之间宽而平。两侧棱明显,中部以上可见窄翼状边缘。种核大小一般为2.1 cm×1.9 cm×1.7 cm(大者可达2.6 cm×2.15 cm×1.63 cm),种核千粒重2 970~3 750 g。出核率23.9%~28.6%。出仁率78.0%。主要分布于山东郯城等地。

'Yuanling'

RLW 1.20. Crown tower-shaped. Leaves triangle or fan-shaped, conspicuous 2-lobed. Fruits ovoid, with cusps or slight emarginated at the apex. The base of stalk scraggly, and circumferentia slight concave. Stalks 3.1cm long; fruit bluish yellow and covered with thin glaicous when ripe. Nuts ovoid, with cusps and two little vascular bundle scars at the base. Raphe conspicuous and wing like from the midst to the apex. Size of nut is 2.1 cm×1.9 cm×1.7 cm (larger one is 2.6 cm×2.15 cm×1.63 cm). 1 000-grain weight of 2 970–3 750 g. PNF 23.9%–28.6% and PKN 78.0% normally. Mainly cultivated in Tancheng, Shandong.

图 3.187
'圆铃'
①江苏邳州嫁接树;
②结果枝生长状态

Fig. 3.187 'Yuanling'
① Grafted tree in Pizhou, Jiangsu, ② Bearing shoots

银杏核用品种——圆子品种群
Yuanzi Group for Nut-producing

图 3.188
'圆铃'
① 种实和种核；
② 种实和叶子；
③ 结果枝蜡叶标本

Fig. 3.188
'Yuanling'
① Fruits and nuts,
② Fruits and leaves,
③ Wax-leaf specimens of bearing shoots

9 '小圆子'

又称'小圆珠'、'小圆头'。核形系数1.38。种实圆形或长圆形，顶部圆钝，顶点稍凹，珠孔迹小。基部平，稍宽圆，四周具纵沟，形成的对角线或"十"字线纵沟明显深长。蒂盘近圆形，稍凹入。种实柄短，平均长约2.6 cm，近蒂盘处稍见粗壮。种实大小约纵径2.69 cm，横径2.29 cm。油胞圆或长圆，凸出种皮之上，并稀疏而均匀地分布于种实中下部，熟时橘黄色或淡黄色，被薄白粉。种核长圆形或近圆形，略扁，两端均钝圆，上下基本一致，先端较基部稍圆。端具小尖，基部束迹迹点细小不够明显，迹点间距较宽，在3.0 mm以上。两侧棱线明显，具厚而狭的翼状边缘，上部稍薄。种核大小为2.2 cm×1.6 cm×1.33 cm，种核千粒重2 200 g。出核率19%～27%。主要分布于江苏吴县，浙江长兴等地。

'Xiao Yuanzi'

'Xiao Yuanzhu', 'Xiao Yuantou'.

RLW 1.38. Fruits ovoid or long ovoid, blunt and round, with small at micropylar, slight depressed the apex, nuts with stria at the base, the diagonal or cross-line stria conspicuous, flattened at the base. Stalks 2.6 cm long. Fruits average 2.69 cm long, 2.29 cm wide. Fruits orange yellow or light yellow and covered with many oil sac and thin glaicous when ripe. Nuts spherical or long spherical, symmetrical, blunt and round at the apex and the base, with cusp and little vascular bundle scars at the base with distance above 3.0 mm, nuts with conspicuous wing-shaped raphe. Size of nut is 2.2 cm×1.6 cm×1.33 cm. 1 000-grain average weight of 2 200 g. PNF 19%–27%. Mainly cultivated in Wuxian, Jiangsu and Changxing, Zhejiang.

图 3.189
'小圆子'
广西桂林海洋乡小圆子结果枝生长状态

Fig. 3.189
'Xiao Yuanzi'
Bearing shoots in Guilin, Guangxi

银杏核用品种——圆子品种群
Yuanzi Group for Nut-producing

图 3.190
'小圆子'
① 种实和种核；
② 种实和叶子；
③ 结果枝蜡叶标本

Fig. 3.190
'Xiao Yuanzi'
① Fruits and nuts,
② Fruits and leaves,
③ Wax-leaf specimens of bearing shoots

171

10 '桐子果'

本品种的种实形状及色泽极似大戟科的油桐种实，故名。核形系数1.18。树势强健。树冠圆锥形，主干挺直，层性明显。长枝上多三角形叶，少数为扇形及截形叶。叶片明显宽大，中裂明显，短枝上中裂较少。种实圆形，先端圆钝，顶微凹，呈"一"字形，具小尖，珠孔迹明显。基部平，蒂盘偏斜，略呈圆形，周缘不整。略凹入。种实柄长约4.1 cm，略弯曲。熟时青黄色，被白粉。种实大小为纵径2.49 cm，横径2.41 cm，单粒种实平均重8.67 g。种核圆形或近圆形。先端圆，顶具小尖。基部稍见平阔，维管束迹迹点大而明显，常一高一低，迹点间距较大，2.9~4.0 mm。两侧棱线明显，自上至下逐步增宽，近尾端处呈窄翼状。种核大小为2.0 cm×1.7 cm×1.18 cm，种核千粒重2 330 g。出核率25.7%。出仁率76%。主产区为广西桂林。

'Tongzi Guo'

Fruits tung oil-shaped. RLW 1.18. Crown spheroidal, main trunk straight. Leaves triangle or fan-shaped, leaves with conspicuous 2-lobed. Fruits ovoid, flattened and round, foveolate, with conspicuous micropylar at the apex, round at the base. The base of stalk round and the circumferentia slight concave. Stalks 4.1cm long and slight oblique, fruits average 2.49 cm long, 2.41 cm wide; fruits green yellow and covered with thicker glaicous when ripe. Nuts ovoid, round and with cusp at the apex. Nuts base flattened with two conspicuous vascular bundle scars with distance 2.9–4.0 mm, nuts with conspicuous raphe, the raphe wing-shaped near to the base. Size of nut is 2.0 cm×1.7 cm×1.18 cm. 1 000-grain average weight of 2 330 g. PNF 25.7% and PKN 76% normally. Mainly cultivated in Guilin, Guangxi.

图 3.191
'桐子果'（后株）
①广西桂林海洋乡30年生实生大树；
②结果枝生长状态

Fig. 3.191
'Tongzi Guo'
① 30 years old tree from seedling in Guilin, Guangxi,
② Bearing shoots

银杏核用品种——圆子品种群
Yuanzi Group for Nut-producing

图 3.192
'桐子果'
① 种实和种核；
② 种实和叶子；
③ 结果枝蜡叶标本

Fig. 3.192
'Tongzi Guo'
① Fruits and nuts,
② Fruits and leaves,
③ Wax-leaf specimens of bearing shoots

11 '皱皮果'

本品种的种实外皮满布皱纹，故名。核形系数1.11。树冠圆锥形。长枝上多三角形叶，扇形叶及截形叶较少，具中裂。叶片较厚，叶色较深。种实椭圆形，先端极尖，基部平。蒂盘椭圆形，种实柄长3.3～3.8 cm，略弯曲。种实大小为纵径2.9 cm，横径2.6 cm，熟时橙黄色，被薄白粉，外种皮满布皱纹。外种皮薄，核易脱出。种核宽椭圆形，有或略有背腹之分。先端圆钝具小尖。基部两束迹迹点明显，间距近，约1.0 mm。两侧棱线明显，中部边棱呈窄翼状。种核大小为1.75 cm×1.57 cm×1.2 cm，种核千粒重2 420 g。出核率达28%，出仁率76%以上。主要分布于广西桂林灵川。

'Zhoupi Guo'

RLW 1.11. Crown spheroidal . Leaves triangle or fan-shaped, with shallow 2-lobed, deep colour. Fruits spherical, with cusps at the apex, flattened at the base, the base of stalk elliptic, stalks slight oblique 3.3–3.8 cm long. Fruits average 2.9 cm long, 2.6 cm wide.Fruits orange yellow and covered with stria and thin glaicous when ripe. Nuts spherical, blunt and round, with cusp at the apex, with conspicuous vascular bundle scars at the base with distance about 1.0 mm; nuts with conspicuous raphe, raphe wing-shaped at midst. Size of nut is 1.75 cm×1.57 cm×1.2 cm. 1 000-grain average weight of 2 420 g. PNF 28% and PKN 76% normally. Only cultivated in Lingchuan, Guangxi.

图 3.193
'皱皮果'
①广西桂林海洋乡30年生实生大树；
②结果枝生长状态

Fig. 3.193
'Zhoupi Guo'
① 30 years old tree from seedling in Guilin, Guangxi,
② Bearing shoots

银杏核用品种——圆子品种群
Yuanzi Group for Nut-producing

图 3.194
'皱皮果'
① 种实和种核；
② 种实和叶子；
③ 结果枝蜡叶标本

Fig. 3.194
'Zhoupi Guo'
① Fruits and nuts,
② Fruits and leaves,
③ Wax-leaf specimens of bearing shoots

3.1.4.3 圆子品种群优良无性系

本节分别对圆子品种群中 6 个优良无性系进行简要介绍，见表 3.4。

3.1.4.3 Clones of Yuanzi Group

Table 3.4 illustrated 6 clones of Yuanzi Group in brief.

表 3.4 圆子品种群优良无性系

Table 3.4 A list of Some Ginkgo Clones of Yuanzi Group

编号 Number	名称 Name	核形系数 RLW	产地 Distribution
1	'道真 5 号' 'Daozhen-5'	1.18	贵州道真 Daozhen, Guizhou
2	'道真 7 号' 'Daozhen-7'	1.23	贵州道真 Daozhen, Guizhou
3	'延安 1 号' 'Yan'an-1'	1.26	陕西延安农业委员会 Yan'an Agriculture Committee, Shanxi
4	'桂林 6 号' 'Guilin-6'	1.27	广西桂林林科所 Guilin Forestry Academy, Guangxi
5	'京山 A23 号' 'Jingshan-A23'	1.28	湖北京山 Jiangshan, Hubei
6	'藤九郎' 'Teng Jiulang'	1.29	日本 Japan

广西桂林海洋乡银杏

Ginkgo in Haiyang, Guilin, Guangxi

银杏核用品种——圆子品种群
Yuanzi Group for Nut-producing

'道真5号' 'Daozhen-5'

图 3.195
'道真5号'
结果枝生长状态

Fig. 3.195
'Daozhen-5'
Bearing shoots

图 3.196
'道真5号'
①种实和种核；
②种实和叶子；
③结果枝蜡叶标本

Fig. 3.196
'Daozhen-5'
① Fruits and nuts,
② Fruits and leaves,
③ Wax-leaf specimens of bearing shoots

'道真7号' 'Daozhen-7'

图 3.197
'道真 7 号'
结果枝生长状态

Fig. 3.197
'Daozhen-7'
Bearing shoots

图 3.198
'道真 7 号'
①结果枝蜡叶标本；
②种实和种核；
③种实和叶子

Fig. 3.198
'Daozhen-7'
① Wax-leaf specimens of bearing shoots,
② Fruits and nuts,
③ Fruits and leaves

银杏核用品种——圆子品种群
Yuanzi Group for Nut-producing

'延安 1 号' 'Yan'an-1'

图 3.199
'延安 1 号'
结果枝生长状态

Fig. 3.199
'Yan'an-1'
Bearing shoots

图 3.200
'延安 1 号'
① 种实和种核；
② 种实和叶子；
③ 结果枝蜡叶标本

Fig. 3.200
'Yan'an-1'
① Fruits and nuts,
② Fruits and leaves,
③ Wax-leaf specimens of bearing shoots

'桂林6号'　'Guilin-6'

图 3.201
'桂林6号'
结果枝生长状态

Fig. 3.201
'Guilin-6'
Bearing shoots

图 3.202
'桂林6号'
①结果枝蜡叶标本；
②种实和种核；
③种实和叶子

Fig. 3.202
'Guilin-6'
① Wax-leaf specimens of bearing shoots,
② Fruits and nuts,
③ Fruits and leaves

银杏核用品种——圆子品种群
Yuanzi Group for Nut-producing

'京山 A23 号' 'Jingshan-A23'

图 3.203
'京山 A23 号'
结果枝生长状态

Fig. 3.203
'Jingshan-A23'
Bearing shoots

图 3.204
'京山 A23 号'
① 种实和种核；
② 种实和叶子；
③ 结果枝蜡叶标本

Fig. 3.204
'Jingshan-A23'
① Fruits and nuts,
② Fruits and leaves,
③ Wax-leaf specimens of bearing shoots

第3章 银杏品种检索表及品种图谱
Chapter 3 Key to Cultivars of *Ginkgo biloba* L. and Illustration of Ginkgo Cultivars

'藤九郎' 'Teng Jiulang'

图 3.205
'藤九郎'
结果枝生长状态

Fig. 3.205
'Teng Jiulang'
Bearing shoots

图 3.206
'藤九郎'
① 结果枝蜡叶标本；
② 种实和种核；
③ 种实和叶子

Fig. 3.206
'Teng Jiulang'
① Wax-leaf specimens of bearing shoots,
② Fruits and nuts,
③ Fruits and leaves

3.2 银杏观赏品种

银杏树体高大，伟岸挺拔，季相分明且有特色，集叶形美、树形美于一身，寿命较长，病虫害少，适宜做庭荫树、行道树或孤植树。银杏观赏品种的开发与利用，对盆景制作、城乡绿化及美化有重要意义。

银杏观赏品种营养器官特性会随环境的变化而发生改变，但经观察，银杏品种成年植株的标准枝中叶形、叶片大小、叶缘、分枝特性、树冠形态等性状在银杏品种内还是比较稳定的，因此，营养器官的特性成为观赏品种分类的主要依据之一。

3.2 Ginkgo Cultivars for Ornamental Purpose

Ginkgo trees are tall and straight, differences in seasons, and they are both beautiful in leaves shape and tree shape, longer life expectancy, fewer pests and diseases, the most suitable for the shade trees, street trees or singular plants trees. The development and use of Ginkgo cultivars for ornamental purpose has great value to bonsai production, rural and urban greening and beautification.

Relatively speaking, vegetative organ characteristics are easy to change while the environment changes, but the leaves shape, leaves size, leaves serrate, branching characters of adult plants and crown form are also stable within a cultivar. They also should be considered as cultivar classification for ornamental purpose.

江苏邳州银杏林
Ginkgo in Pizhou, Jiangsu

3.2.1 银杏观赏品种分类检索表

银杏观赏品种分类检索表

1. 枝条下垂
  2. 树形宽展
    3. 枝条水平伸展，顶端下垂..1.'垂枝银杏'
    3. 枝条水平伸展，很少直立向上..2.'展冠银杏'
  2. 树形紧凑..3.'塔状银杏'
1. 枝条直立向上生长
  4. 树体矮小，灌木
    5. 叶平展，不卷曲
      6. 叶先端具 2 裂刻
        7. 叶片宽阔，叶先端波状浅裂
          8. 叶色变异，春夏季叶是黄色，夏末秋初变为淡绿色（幼叶除外），直到落叶之前又变为黄色..4.'万年金'
          8. 叶色不具变异，春夏秋季全为绿色，落叶之前变为黄色.............5.'玉蝴蝶'
        7. 叶片狭窄，叶先端深裂至底部..6.'狭叶银杏'
      6. 叶掌状深裂至叶片的 1/2，裂刻数 5～7 个..............................7.'掌状银杏'
    5. 叶管状，卷曲..8.'筒叶银杏'
  4. 树体高大，乔木
    9. 树冠卵圆形，密集
      10. 树形为金字塔形
        11. 枝条稀疏..9.'金秋-1'
        11. 枝条密集..10.'费尔蒙特'
      10. 树形为圆柱形
        12. 基部比上部稍宽..11.'金兵普林斯顿'
        12. 基部与上部等宽..12.'五月田野'
    9. 树冠圆球形，宽展..13.'金球'

韩国银杏景观路
Ginkgo Road in Korea

3.2.1 Key to Ginkgo Cultivars for Ornamental Purpose

Key to Ginkgo Cultivars for Ornamental Purpose

1. Branches pendulous
   2. Crown spreading, broad and wide
      3. Branches horizontal spreading, with tips pendulous .... 1. 'Chuizhi Ginkgo'
      3. Branches horizontal spreading, rarely upright .... 2. 'Zhanguan Ginkgo'
   2. Crown densely compacted .... 3. 'Tazhuang Ginkgo'
1. Branches upright
   4. Shrubs
      5. Leaves blade explanate
         6. Leaves blade flabellate, apex 2-lobed
            7. Leaves blade broad, margin loosely wavy
               8. Leaves color character variance, yellow in spring and summer, jade-green in end summer and the early autumn (new leave preclusive), yellow again in defoliation .... 4. 'Wannian Jin'
               8. Leaves color character is stable, green in spring and summer, yellow before defoliation .... 5. 'Yu Hudie'
            7. Leaves blade narrow, deeply divided at the base .... 6. 'Xiaye Ginkgo'
         6. Leaes blade palmatifid, 5-7segments .... 7. 'Zhangzhuang Ginkgo'
      5. Leaves blade tubular, involute .... 8. 'Tongye Ginkgo'
   4. Trees
      9. Crown ovoid, compacted
         10. Crown broadly pyramidal
            11. Branches sparsity .... 9. 'Jin Qiu-1'
            11. Branches compacted .... 10. 'Feiermengte'
         10. Crown nearly columuar
            12. Crown slightly broad at the base .... 11. 'Jinbing Pulinsidun'
            12. Crown from the base to top the equivalent width ...... 12. 'Wuyue Tianye'
      9. Crown rounded, broadening .... 13. 'Jin Qiu'

3.2.2 银杏观赏品种描述

本节分别对银杏13个主要观赏品种进行详细介绍

3.2.2 Illustration of Ornamental Cultivars

Illustrated 13 cultivars for Ornamental Purpose of Ginkgo in detail.

1 '垂枝银杏'

雌株。中等生长速度，小树有宽阔半圆形的树冠。不规则的水平生长习性产生宽阔、扭曲的特征。枝条或多或少下垂（垂枝），生长慢，装饰性强。这种树的下垂特征并没有同其他下垂树种的定义那么精确，但这种树确实比其他树种更具有水平生长的习性。在生长初期与'展冠银杏'相似，但随着枝条生长而下垂。也可以在生长早期进行修剪，但如果任其发展将更具观赏价值。深绿色叶，独特叶形，极好看的金黄色落叶。

'Chuizhi Ginkgo'

'Pendula'
Female. Medium growing, small tree with a broad hemispherical crown. Irregular, horizontal growth habit forming a broad twisted character. Branches more or less pendulous 'weeping', slow growing, decorative. The weeping character of this tree is not as clearly defined as with other weeping trees. The branches do have a more horizontal character than other varieties. Similar with horizontalis in its early stage but its growth is pendulous. It may also be trimmed in its early stage but a fascinating plant will be formed if left to itself. Dark-green, uniquely-shaped leaves adorn this desirable garden tree.

图 3.207
广西桂林海洋乡'垂枝银杏'

Fig. 3.207
'Chuizhi Ginkgo' in Haiyang, Guilin, Guangxi

2 '展冠银杏'

雌株，树冠宽大。高大宽展、散布形，很少直立形。很多侧枝。一个优美的侧俯形。长长的侧枝，接近地面，产生散布垂枝的效果。靠近墙壁或假山栽植，其层叠的枝条生动美观。如名字所示，这种树平直的生长，尽管其可以修建成直立形，但过段时间就会像一把伞。

'Zhanguan Ginkgo'

'Horizontalis'

Female. Wide crown, tall and wide forms, less upright. Many side-branches. A very graceful, prostrate form. Long arching branches close to the ground creating a spreading weeping effect. Plant nearby a low wall or rockery where its cascading branches can tumble over for dramatic effect. As the name after the habit of this tree is remarkably flat growing although it may be trimmed upright initially, the crown of this plant looks like a huge umbrella. A strikingly shape will be appeared quickly after raised in pots.

图 3.208
广西桂林海洋乡 '展冠银杏'

**Fig. 3.208
'Zhanguan Ginkgo' in Haiyang, Guilin, Guangxi**

3 '塔状银杏'

雌株，矮小，大约1 m高。更加紧凑的圆形的树形，水平枝条在顶端下垂，造成枝条或多或少的下垂，应该嫁接在1.5 m高的树桩上获得大树。扇形叶具有深绿色颜色。好看的金黄色落叶，是紧凑且具有吸引力的矮小银杏。

'Tazhuang Ginkgo'

'Mariken'

Female. Dwarf, compact, about 1 m high, with a tight-rounded habit. Horizonal branching with pendulous tips made branches a little pendulous in habit. It should be grafted on about 1.5 m stock for getting a taller plant. Dark green color appears in scallop-edged leaves. The fall leaves has golden color.

图 3.209 广西桂林海洋乡'塔状银杏'

Fig. 3.209 'Tazhuang Ginkgo' in Haiyang, Guilin, Guangxi

4 '万年金'

雌株，亲本来源于湖北安陆市王义贞镇唐僧村，胸径43.00，冠幅18.90 m×17.50 m。在生长期内有一根大枝叶色为黄色，其他枝条上的叶色均为绿色。连续3年观察发现从萌芽开始，一直到7月底，该枝条上的银杏叶均为黄色，8月份以后，除新发的幼叶为黄色外，成熟的黄色叶逐渐转为淡绿，11月份以后又变为黄色，性状稳定。从该枝条下剪下接穗，发现嫁接后的银杏苗木从第二年开始叶色变化情况与亲本枝条上的叶色变化一致，遗传性状稳定。测定结果表明，'万年金'银杏叶片中的类胡萝卜素含量/叶绿素总量的比值在整个生长期内均较一般银杏品种的叶片高。

'Wannian Jin'

Female. The living specimen grows in Tangseng of Wangyizhen, Anlu, Hubei, dbh 43.00 cm, and crown 18.90 m × 17.50 m. The leaves on one of big branches are always yellow during growing period, others are green. From sprouting to end of July, the leaves on this branch are always yellow, the leaves (immature leave preclusive) will turn into jade-green, turn back to yellow again in November, and the characters is stable. After grafting, the genetic character is always stable. The determination results show that ratio of carotenoid content and total chlorophyll in leaves of 'Wannian Jin' is always higher than other ginkgo varieties during growing period.

图 3.210
湖北安陆'万年金'

Fig. 3.210 'Wannian Jin' in Anlu, Hubei

5 '玉蝴蝶'

雄株。灌木型，花瓶形状，半矮生，达3 m。浓密的深绿色叶片簇生，似一只只蝴蝶。秋天黄色落叶似奶油。

'Yu Hudie'

'Jade Butterfly'

Male. Shrubby outline, vase shaped, semi dwarf, about 3 m high. Dense dark green foliage clumps, similar to a butterfly. The autumn foliage looks like cream.

图 3.211 加拿大列治文'玉蝴蝶'

Fig. 3.211 'Yu Hudie' in Richmond, Canada

6 '狭叶银杏'

雌株。灌木型，大概12 m高，7 m宽。直立圆形的外形，生长缓慢。分枝浓密对称的品种，生长至12.50 m × 9.40 m，宽阔的圆形树冠是良好的遮荫树。叶片狭窄，深裂至底端，叶片下垂（像一半展开的扇子或袋袋裤）。

'Xiaye Ginkgo'

'Saratoga'

Female. Shrubby plant with a crown reaches to 12.50 m × 9.40 m. Erect, rounded appearance with slow growth. This variety with dense symmetrical branching, especially bread for a broad canopy making it the best of the genus for a shade tree. Narrow leaves deeply split at ends, pendulous (resembles small, half open fans or looks like clowns baggy trouserous).

7 '掌状银杏'

雌株。本品种最大的特点是叶大、多裂刻，叶形呈掌状深裂至叶片的1/2，裂刻数5～7。从形态上看，叶面非裂部位宽仅2～3 cm，已失去银杏扇形叶的特征。

'Zhangzhuang Ginkgo'

'Dissected Ginkgo'

Female. Large leaves, with palmatifid, 5–7 segments. From the morphological point of view, it has no fan-shaped character.

图 3.212
'掌状银杏'

Fig. 3.212
'Zhangzhuang Ginkgo'

8 '筒叶银杏'

雌性无性系。树高3 m，灌木，细弱小枝，有两种类型的叶片。在成熟茎干上是卷曲的，类似管子或小喇叭，并由此得名，而在幼嫩枝条上是正常的扇形叶片，但是有锯齿轮廓。随着树的成熟，大多数叶片成为喇叭状。

'Tongye Ginkgo'

'Tubifolia', 'Tubiformis'

Female. Shrubby plants about 3 m high. A very unusual shrub by Ginkgo produces lots of thin twiggy growth that is adorned by two different types of leaves. On mature stems the leaves are fused, rolled up like a tube or a trumpet, hence 'Tubifolia' whereas younger stems produce regular Ginkgo leaves but with a jagged outline. As the tree matures the majority of the leaves has a trumpet shape.

图 3.213
'筒叶银杏'

Fig. 3.213
'Tongye Ginkgo'

9 '金秋-1'

雌株。直立对称的树形，改良的宽阔散布生长习性，宽阔的金字塔型，更加紧凑，成熟时达到15.24 m高，9.14 m宽。如名字所示，秋天变成很好看的金秋色，且比其他品种持续的时间长。

'Jin Qiu-1'

'Autumn Gold'

Female with an upright, symmetrical form, modified a broad spreading growth habit. It has a broadly pyramidal growth form at maturity and tends to be a little more compact than the species, attaining a size of approximately 15.24 m high and 9.14 m wide. As the name suggests the leaves turn a rich gold color in autumn and remain on the branches longer than most other varieties.

图 3.214
广西桂林海洋乡 '金秋-1'

**Fig. 3.214
'Jin Qiu-1' in Haiyang, Guilin, Guangxi**

10 '费尔蒙特'

健壮的雄株，具有较好的向上生长的习性。纤细型，密集金字塔形树冠，直立的主干。生长迅速，成熟时达到21.34 m高，9.14 m宽。叶形较大。

'Feiermengte'

'Fairmount'

Male with superior upright habit. Crown pyramidal, slender and narrow, dense. It can reach to 21.34 m high and 9.14 m spread. Larger leaves.

11 '金兵普林斯顿'

直立向上生长的雄株，生长缓慢。树干近圆柱形，基部稍有些宽。直立生长的圆锥形提供了正常的焦点，大的装饰性的叶片。金黄色落叶。是'Fastigiata'的改良品种。名字来源于普林斯顿坟场。

'Jinbing Pulinsidun'

'Princeton Sentry'

Male. Trunk upright, cylindric, slight wide at the base. Upright conical form gives very formal focal point. Big decorative leaves with golden yellow color in fall. Name derived from Princeton Cemetary.

12 '五月田野'

雄株，树形柱状。非常高，狭窄，向上，生长缓慢，成熟时达到9.78 m高，3.65 m宽。枝条较短。

'Wuyue Tianye'

'Mayfield'

Male clone with a columnar form. Extremely tall, narrow, upright, and slow growing. Tight short branches. It reaches to 9.78 m high with a 3.65 m spread.

13 '金球'

雌株。幼小的树木有完整树冠，成熟时是宽阔、圆形。树木有不寻常的浓密分枝。秋天有壮观的金黄色叶子。比其他树种生长较快。

'Jin Qiu'

'Golden Globe'

Female. Young plants have full crowns with a broad, rounded top when it matures. Trees are unusually have densely branches. Spectacular golden foliage in the fall. Compared to many other varieties, it grows faster.

图 3.215
韩国'金球'

Fig. 3.215
'Jin Qiu' in Korea

3.3 银杏叶用优良无性系

银杏叶具有很高的药用价值，银杏叶中主要化学活性物质黄酮类化合物和萜内酯类化合物，具有捕获自由基、抑制血小板活化因子（PAF），促进血液循环及脑代谢的功能。因此，银杏叶可用于治疗多种疾病，同时还可作为添加剂用于保健食品和化妆品的生产。

银杏叶用品种以叶大、高产、内黄酮和萜内酯含量高为主要选种目标，这里重点介绍近年来我国林业研究人员初选出的一些优良叶用无性系。

分别对 11 个银杏叶用优良无性系进行简要介绍。

江苏泰兴银杏与桑树复合经营
Agroforestry Management of Ginkgo and Mulberry in Taixing, Jiangsu

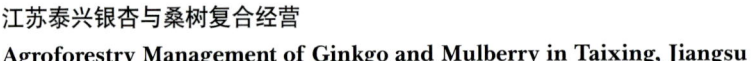

3.3 Ginkgo Clones for Leaf-producing

Ginkgo leaves have high medicinal value, the main chemical active material in Ginkgo leaves flavonoids and terpene lactones capture compounds with free radicals, inhibiting platelet-activating factor (PAF), promote blood circulation and metabolism in the brain function. It not only use for the treatment of many disease, but also as an additive for health food and cosmetics production. Flavonoids and terpene lactones in Ginkgo leaves are significant difference, large leaves and high yield, good quality as the main target selection for the leaf-producing Ginkgo.

Illustrated 11 Ginkgo clones for leaf-producing in brief.

1 '黄酮 F-1 号'

雄株。原株来自山东，树高 15.9 m，胸径 72 cm，树龄 100 年。标准叶菱形，边缘浅波状，基部楔形。叶裂刻 1 个，长×宽为 5.5 cm×0.8 cm。油胞稀、团状、较大，分布在叶子中下部，叶绿色。单株新梢数 37.8 个，叶数 1169 片，枝条总长度 13.1 m。当年新梢长 56.83 cm，粗 0.98 cm，每梢叶数 41.5 个。叶面积系数 2.32。

莱州试验点连续 3 年黄酮苷测定结果为，2.96%（1 年）、2.33%（2 年）和 1.96%（3 年）。郯城分别达 2.59%、1.77% 和 1.57%。药乡分别为 2.85%、1.90% 和 2.38%。属高黄酮苷良种。内酯含量 0.134 2%，其中 BB 0.026 2%，GJ 0.002 8%，GC 0.021 2%，GA 0.029 9% 和 GB 0.018%。产量性状接后 1 年产鲜叶 0.042 kg/株，2 年 0.335 kg/株，3 年 0.543 kg/株（黄酮苷采用 754 分光光度法测，萜内酯采用 HPLC 法测据）。

'Huangtong-F1'

Male. Mother tree from in Shandong. The tree is 15.90 m tall and 72.00 cm in d.b.h, 100 years old. Leaves green, rhombic, margin sinuolate, wedged at the base. One lobe, long 5.50 cm and wide 0.80 cm. Ocellus sparse, cluster structure, distributed in middle-lower part of leaves. The tree with 37.80 current shoots and 1169 leaves, the total length of branches is 13.10 m. Current shoots long 56.83 cm and with diameter 0.98 cm, each current shoot with 41.5 leaves. The leaf area index is 2.32.

The experimental results of flavonoid glycosides content at different experimental sites show that with flavonoid glycosides content 2.96% (1st year), 2.33% (2nd year) and 1.96% (3rd year) at Laizhou and 2.59% (1st year), 1.77% (2nd year) and 1.57% (3rd year) at Tancheng, 2.85% (1st year), 1.90%(2nd year) and 2.38%(3rd year) at Yaoxiang. Huang tong F-1 is a good variety with high flavonoid glycosides content. The mean value of lactone content is 0.134 2%, BB 0.026 2%, GJ 0.002 8%, GC 0.021 2%, GA 0.029 9% and GB 0.018%. The trait of fresh leaf yield shows mean yields of fresh leaf after graft are 0.042 kg (1st year), 0.335 kg (2nd year) and 0.543 kg (3rd year). Flavonoid glycoside determined by 754 Spectrophotometry and terpene lactone determinated by HPLC (High Performance Liquid Chromatography).

2 '黄酮 F-2 号'

雌株。原株产于江苏，定型叶叶形为半圆形、叶缘波状、基部截形，裂刻 1 个，长×宽为 2.8 cm×0.8 cm。油胞稀呈放射状分布在叶子上部。叶绿色。单株新梢数 23.7 个，叶数 910 片，枝总长 9.37 m。当年新梢长 44.2 cm，粗 0.96 cm，每梢叶数 55.7 片。叶面积系数 4.82。郯城试验点三年黄酮测定分别为 2.61%、2.72%、2.43%，莱州分别为 2.01%、1.89%、2.19%，药乡林场分别达 2.59%、3.03% 和 2.81%。属高黄酮苷品种。内酯为 0.28%，其中 BB 0.020 5%，GJ 0.010 6%，GC 0.037 8%，GA 0.165 2%，GB 为 0.047 2%。产量，接后 1～3 年分别株产鲜叶 0.124 kg、0.425 kg、0.535 kg（黄酮苷采用 754 分光光度法测，萜内酯采用 HPLC 法测据）。

'Huangtong-F2'

Female. Mother tree in Jiangsu, leaves green, semicircular, margin undulance, truncated at the base. One lobe, long 2.80 cm and wide 0.80 cm. Ocellus radial distributed in upper part of leaf. The tree with 23.70 current shoots and 910 leaves. Current shoots long 44.20 cm and with diameter 0.96 cm, each current shoot with 55.70 leaves. The leaf area index is 4.82. The experimental results of flavonoid glycosides content at different experimental sites show that with flavonoid glycosides content 2.61% (1st year), 2.72% (2nd year) and 2.43% (3rd year) at Tancheng and 2.01% (1st year), 1.89% (2nd year) and 2.19% (3rd year) at Laizhou, 2.59% (1st year), 3.03% (2nd year) and 2.81% (3rd year) at Yaoxiang. Huang tong F-2 is a good variety with high flavonoid glycosides content. The mean value of lactone content is 0.28%, BB 0.020 5%, GJ 0.010 6%, GC 0.037 8%, GA 0.165 2% and GB 0.047 2%. The trait of fresh leaf yield shows mean yields of fresh leaf after graft are 0.124 kg (1st year), 0.425 kg (2nd year) and 0.535 kg (3rd year). Flavonoid glycoside determined by 754 Spectrophotometry and terpene lactone determined by HPLC (High Performance Liquid Chromatography).

3 '黄酮 F-3 号'

雄株。原株来自山东，由树高 15.7 m，胸径 62 cm，树龄 80 年。定型叶菱形，叶缘浅波状，基部楔形。大多具一个裂刻，长 × 宽为 2.5 cm×1.0 cm。油胞较稀，团状分布在叶子中下部，叶色浓绿。单株新梢数 35 个，叶数 1005 片，枝总长 23.32 m。当年新梢长 59.00 cm，粗 1.06 cm，每梢叶数 52 片。叶面积指数 1.76。

莱州点连续 3 年黄酮测定为 2.23%（1 年）、1.80%（2 年）和 1.24%（3 年）；郯城分别为 2.87%、2.69% 和 1.55%，药乡林场分别为 2.39%、3.86% 和 3.12%。属于高黄酮苷良种。内酯含量 0.106%，其中 GJ 0.009 1%，BB 0.008 1%，GC 0.012 3%，GA 0.044 2%，GB 0.032 2%，产量性状，接后 1 年株产鲜叶 0.29 kg、2 年 0.54 kg、3 年 0.605 kg（黄酮苷采用 754 分光光度法测，萜内酯采用 HPLC 法测据）。

'Huangtong-F3'

Male. Mother tree in Shandong. The tree is 15.7 m tall and 62.00 cm in d.b.h, 80 years old. Leave deep green, rhombic, margin sinuolate, wedged at the base. One lobe, long 2.50 cm and wide 1.00 cm. Ocellus sparse, cluster structure, distributed in middle-lower part of leaf. The tree with 35 current shoots and 1005 leaves, the total length of branches is 23.32 m. Current shoots long 59.00 cm and with diameter 1.06 cm, each current with 52 leaves. The leaves area index is 1.76.

The experimental results of flavonoid glycosides content at different experimental sites show that with flavonoid glycosides content 2.23% (1st year), 1.80% (2nd year) and 1.24% (3rd year) at Laizhou and 2.87% (1st year), 2.69% (2nd year) and 1.55% (3rd year) at Tancheng, 2.39% (1st year), 3.86% (2nd year) and 3.12% (3rd year) at Yaoxiang. Huang tong F-3 is a good variety with high flavonoid glycosides content. The mean value of lactone content is 0.106%, BB 0.008 1%, GJ 0.009 1%, GC 0.012 3%, GA 0.044 2% and GB 0.032 2%. The trait of fresh leaf yield shows mean yields of fresh leaf after graft are 0.29 kg (1st year), 0.54 kg (2nd year) and 0.605 kg (3rd year). Flavonoid glycoside determined by 754 Spectrophotometry and terpene lactones determined by HPLC (High Performance Liquid Chromatography).

4 '内酯 T-5 号'

雌株。原株来自山东，实生树，树龄约 25 年生，树高 9.5 m，胸径 25 cm。标准叶半圆形、叶缘波状、基部截形、1 个裂刻，长 × 宽为 7.8 cm×2.5 cm。油胞密呈圆点状、放射状分布在叶子中上部，叶色浓绿。单株新梢数 22 个，叶数 744 片，枝总长 13.12 m。单梢长 71 cm，粗 1.55 cm，每梢叶数 57 片。叶面积指数 7.05。

萜内酯总量（HPLC）达 0.405 8%，其中 GJ 0.032 5%、GC 0.061 8%、GA 0.240 4%、GB 0.040 4%、BB 0.030 9%。属高内酯无性系。黄酮含量 2.58%（1 年）、2.13%（2 年）和 1.13%（3 年）。1～3 年生单株产叶量分别为 0.112 kg、0.5 kg 和 0.76 kg。T-5 为高内酯、高黄酮及高产无性系（黄酮苷采用 754 分光光度法测，萜内酯采用 HPLC 法测据）。

'Neizhi-T5'

Female. Mother tree in Shandong, the tree is 9.5 m tall and 25 cm in d.b.h, 25 years old. Leavest deep green, semicircular, margin undulance, truncated at the base. One lobe, long 7.8 cm and wide 2.5 cm. Ocellus dense, punctate, radial distributed at upper part of leaf.

The tree with 22 current shoots and 744 leaves, the total length of branches is 13.12 m. Current shoots long 71cm and with diameter 1.55 cm, each current with 57 leaves. The leaves area index is 7.05.

The flavonoid glycosides content is 2.58% (1st year), 2.13% (2nd year) and 1.13% (3rd year). The mean value of lactone content is 0.405 8%, BB 0.030 9%, GJ 0.032 5%,GC 0.061 8%,GA 0.240 4% and GB 0.040 4%. The mean yields of fresh leaf of grafting trees are 0.112 kg (1st year), 0.5 kg (2nd year) and 0.76 kg (3rd year). Nei zhi T-5 is a good variety with high lactone content and flavonoid glycosides content.Flavonoid glycoside determined by 754 Spectrophotometry and terpene lactones determined by HPLC (High Performance Liquid Chromatography).

5 '内酯 T-6 号'

雌株。原株来自山东，树龄约 400 年生，实生。树高 21 m，胸径 0.94 m。冠幅

12.5 m×14.0 m，枝下高 2.58 m，枝散生，7 个主枝，海拔约 151 m。长枝上的定型叶宽扇形，叶缘浅波状，基部楔形，具 1 个裂刻，长 × 宽为 7.2 cm×2.4 cm。油胞极稀，星点状，较小，放射状分布于叶子外缘。叶色浅绿至深绿。单株叶数 553 片，新梢数 13 个，枝总长 556 cm，冠幅 105 cm×59 cm，LAI 4.57。单梢长 53 cm，粗 1.05 cm，每梢叶数 50 片。萜内酯总量（HPLC）达 0.358 4%，其中 GJ 0.054 7%，GC 0.109 7%，GA 0.073%，GB 0.067 7%，BB 0.053 3%。属高内酯无性系。黄酮含量 1.76%（1 年）、1.53%（2 年）和 1.47%（3 年）。1～3 年生单株产叶量分别为 0.017 kg、0.258 kg 和 0.43 kg（黄酮苷采用 754 分光光度法测，萜内酯采用 HPLC 法测据）。

'Neizhi-T6'

Female. Mother tree in Shandong, the tree is 21.00 m tall and 94.00 cm in d.b.h, 400 years old, crown width are 12.50 m and 14.00 m, pole height 2.58 m, grows at the sea level 151.00 m.

Leaves light green to deep green, fan-shaped, margin sinuolate, wedged at the base. One lobe, long 7.20 cm and wide 2.40 cm. Ocellus punctate, radial distributed at outer margin of leaves. The grafting tree with 13 current shoots and 553 leaves, the total length of branches is 556 cm. Current shoots long 53cm and with diameter 1.05 cm, each current with 50 leaves. Crown width are 105.00 cm× 59.00 cm, the leaf area index is 4.57.

The flavonoid glycosides content is 1.76% (1st year), 1.53% (2nd year) and 1.47% (3rd year). The mean value of lactone content is 0.358 4%, BB 0.053 3%, GJ 0.054 7%, GC 0.109 7%, GA 0.073% and GB 0.067 7%. The mean yields of fresh leaf of grafting trees are 0.017 kg (1st year), 0.258 kg (2nd year) and 0.43 kg (3rd year). Nei zhi T-6 is a good variety with high lactone content. Flavonoid glycoside determined by 754 Spectrophotometry and terpene lactone determined by HPLC (High Performance Liquid Chromatography).

6 '内酯 GB-5 号'

雌株。原株来自山东，实生大树，树龄约 400 年，树高 10.5 m，胸径 1.16 m。冠幅 9.3 m×1.75 m，枝下高 3.8 m，4 大主枝，生长旺盛。长枝上的定型叶呈半圆形，边缘浅波状，基部楔形，具 1 个裂刻，长 × 宽为 2.2 cm×0.4 cm。油胞较稀，为长椭圆形，较大，点状分布在整个叶面。叶淡绿色。单株梢数 55 个，叶数 1 432 片，枝总长 16.58 m。冠幅 156 cm×120 cm，LAI 为 4.74。单梢长 38.5 cm，粗 0.9 cm，每梢叶数 49 片。

萜内酯总量（HPLC）达 0.265 4%，其中 GB 高达 0.089 2%，GJ 0.019%，GC 0.027 1%，GA 0.105 4%，BB 0.024 7%。属高 GB 无性系。黄酮含量 1.55%（1 年）、1.22%（2 年）和 1.04%（3 年）。产量性状，1～3 年生单株产叶量分别达 0.172 kg、0.36 kg 和 0.528 kg（黄酮苷采用 754 分光光度法测，萜内酯采用 HPLC 法测据）。

'Neizhi-GB5'

Female. Mother tree in Shandong, the tree is 10.50 m tall and 116.00 cm in d.b.h, 400 years old, crown width are 9.30 m × 1.75 m, pole height 3.80 m, 4 scaffold limbs.Leaves light green, semicircular, margin sinuolate, wedged at the base. One lobe, long 2.20 cm and wide 0.40 cm. Ocellus sparse, long oval, point shape distributed at leaves. The grafting tree with 55 current shoots and 1 432 leaves, the total length of branches is 16.58 m. Current shoots long 38.5 cm and with diameter 0.90 cm, each current with 49 leaves. Crown width are 156.00 cm × 120.00 cm, the leaf area index is 4.74.The flavonoid glycosides content is 1.55% (1 year), 1.22% (2 years) and 1.04% (3 years). The mean value of lactone content is 0.265 4%, BB 0.024 7%, GJ 0.019 7%, GC 0.027 1%, GA 0.105 4% and GB 0.089 2%. The mean yields of fresh leaf of grafting trees are 0.172 kg (1st year), 0.36 kg (2nd year) and 0.528 kg (3rd year). Nei zhi GB-5 is a good GB clone with high lactone content. Flavonoid glycoside determined by 754 Spectrophotometry and terpene lactone determined by HPLC (High Performance Liquid Chromatography).

7 '高优 Y-2 号'

雄株。原株来自山东,树龄 100 年,树高 18.1 m,胸径 0.69 m,冠幅 9.7 m × 8.8 m,枝下高 4.0 m,主枝数 10 个,海拔 440 m,生长旺盛。叶子半圆形,边缘浅波状或全缘。雄花散粉期 4 月 17 日,盛期 4 月 20 日,末期 4 月 24 日,共计 9 天时间。1～4 年生枝开花率 6.4%、91.0%、96.7% 和 89.2%。短枝粗 0.57 cm,长 0.37 cm,每个短枝上有小孢子叶球 3.1～4.5 个,出粉率 3.91%。每短枝叶数 4.7,节间长 2.19 cm,短枝数每米长枝 42.63,叶数每米长枝 200 片,叶面积每米长枝 7 806 cm^2,叶鲜重每米长枝 255.7 g,叶干重每米长枝 60.36 g。每个短枝上的叶面积 202.16 cm^2,叶鲜重 7.24 g,叶干重 1.71 g。嫁接 3 年生苗株产鲜叶 0.59 kg,总黄酮含量 1.96%,内酯 0.212%,其中 GJ 0.009%、GC 0.046%、GA 0.038%、GB 0.079% 和 BB 0.04%(黄酮苷采用 754 分光光度法测,萜内酯采用 HPLC 法测据)。

'Gaoyou-Y2'

Male. Mother tree in Shandong, the tree is 18.10 m tall and 69.00 cm in d.b.h, 100 years old, crown width are 9.70 m × 8.80 m, pole height 4.00 m, 10 scaffold limbs, grows at the sea level 440.00 m. Leaves semicircular, margin sinuolate. Days of pollen maturation are 9 days, April 17th to 24th . The flowering rate of different shoots are 6.4% (one year), 91.0% (two years old), 96.7% (three years old) and 89.2% (four years old). Short branchlets long 0.37 cm and with diameter 0.57 cm, each short branchlets with staminate strobilus range 3.10 to 4.50 and 4.7 leaves, the rate of pollen production is 3.91%. The 3 years old grafting trees with flavonoid glycosides content 1.96% , the mean value of lactone content 0.212%, BB 0.04%, GJ 0.009%, GC 0.046%, GA 0.038% and GB 0.079%. The mean yields of fresh leaf of 3 years old is 0.59 kg. Flavonoid glycoside determined by 754 Spectrophotometry and terpene lactone determined by HPLC (High Performance Liquid Chromatography).

8 '丰产 Y-8 号'

雌株。原株产于江苏，生长旺盛，发枝力强。标准叶半圆形，叶缘波状，基部截形，具一个裂刻，长 × 宽为 3.8 cm × 2.0 cm。油胞稀，圆点状，较小，分布在叶基，叶绿色。接后 3 年短枝上的叶，叶长 6.2 cm，叶宽 9.9 cm，叶柄长 6.5 cm，叶面积 33.24 cm^2，鲜重 1.06 g，干重 0.33 g，含水量 68.66%，基线夹角 138°；长枝上的叶分别为叶长 7.2 cm，叶宽 11.9 cm，叶柄长 7.2 cm，叶面积 58.68 cm^2，鲜重 2.31 g，干重 0.79 g，含水量 65.78% 和基线夹角 153°。单株新梢数 33.66 个，每株叶数 1 000 片，枝条总长 9.61 m。单梢长 48.83 cm，粗 1.10 cm，每梢叶数 34.66 片。叶面积指数 4.91。接后当年株产鲜叶 0.035 kg，2 年 0.269 kg，其中短枝占 16.52%，长枝占 83.48%；接后 3 年株产鲜叶 0.499 kg，其中短枝占 35.38%，长枝占 64.62%。总黄酮苷 1 年生苗 2.65%，2 年 1.56%，3 年 1.28%。总萜内酯 0.1145%，其中 BB 0.007 2%，GJ 0.008 7%，GC 0.024 2%，GA 0.060 9%，GB 0.013 5%（黄酮苷采用 754 分光光度法测，萜内酯采用 HPLC 法测据）。

'Fengchan-Y8'

Female. Mother tree in Jiangsu, leaves green, semicircular, margin undulance, truncated at the base. One lobe, long 3.80 cm and wide 2.00 cm. Ocellus sparse, punctate, distributed at the base of leaf. The 3 years old grafting tree, leaves on short branchlets, long 6.20 cm, wide 9.90 cm, leave stalks long 6.50 cm, leaves area 33.24 cm^2, fresh weight 1.06 g, dry weight 0.33 g, water content 68.66%; leaves on long branchlets, long 7.20 cm, wide 11.90 cm, leave stalk long 7.20 cm, leaves area 58.68 cm^2, fresh weight 2.31 g, dry weight 0.79 g, water content 65.78%.The each grafting tree with 33.66 current shoots and 1 000 leaves, the total length of branches is 9.61 m. Current shoots long 48.83 cm and with diameter 1.10 cm, each current with 34.66 leaves, the leaf area index is 4.91. The mean yields of fresh leaf of grafting trees are 0.035 kg (1st year), 0.269 kg (2nd year), which short branchlets yields 16.52% of it and long branchlets yields 83.48% of it; and 0.499 kg (3rd year), which short branchlets yields 35.38% of it and long branchlets yields 64.62% of it. The flavonoid glycosides content is 2.65% (1st year), 1.56% (2nd year) and 1.28% (3rd year). The mean value of lactone content is 0.114 5%, BB 0.007 2%, GJ 0.008 7%, GC 0.024 2%, GA 0.060 9% and GB 0.013 5%. Flavonoid glycoside determined by 754 Spectrophotometry and terpene lactone determined by HPLC (High Performance Liquid Chromatography).

9 '丰产 Y-6 号'

雄株，原产广西。树龄约 50 年，树高 15.7 m，胸径 0.56 m。定型叶心形，全缘，1 个裂刻将叶子平分为两部分。裂刻长 × 宽为 3.8 cm×0.5 cm。油胞稀而小，放射状分布在叶子中下部。叶色浓绿，有别于其他品种，叶形较独特，同时可以作为观赏品种。1 年生长枝叶明显大，且长短枝叶差异较小，短枝上的叶比一般品种大而均匀，叶柄较粗。叶面积指数 1.88。接后当年株产鲜叶 0.165 kg，2 年 0.431 kg，3 年 0.783 kg。短枝叶占 37.73%，长枝占 62.27%，差异较其他品种小。当年生叶黄酮总量达 1.786%，内酯总量 0.077%，其中 GJ 0.009 4%、GC 0.019%、GA 0.021 8%、GB 0.011 7%、BB 0.014 9%。该无性系属高产观赏兼优品种（黄酮苷采用 754 分光光度法测,萜内酯采用 HPLC 法测定）。

'Fengchan-Y6'

Male. Mother tree in Guangxi, the tree is 15.70 m tall and 56.00 cm in d.b.h, 50 years old. Leaves deep green, heart-shaped, entire margin. One deep lobe divides the leaves into two parts, long 3.80 cm and wide 0.50 cm. Ocellus sparse, radial distributed in lower part of leaf. The mean yields of fresh leaf of grafting trees are 0.165 kg (1st year), 0.431 kg (2nd year) and 0.783 kg (3rd year), which short branchlets yields 37.73% of it and long branchlets yields 62.27% of it. The current leaves, flavonoid glycosides content is 1.786% and the lactone content is 0.077%, BB 0.014 9%, GJ 0.009 4%,GC 0.019%, GA 0.021 8% and GB 0.011 7%. Flavonoid glycoside determined by 754 Spectrophotometry and terpene lactone determined by HPLC (High Performance Liquid Chromatography).

10 '丰产 Y-3 号'

雄株，原产山东。树龄约 100 年，树高 21.2 m，胸径 0.65 m。实生树，主干明显。冠幅 7.5 m × 10.4 m，枝下高 5.4 m，海拔 200 m 左右。生长旺盛。定型叶宽扇形，浅波状，基部为楔形，大多具 1 个裂刻，长 × 宽为 8.2 cm × 3.3 cm，叶子烘干后油胞明显而密，油胞为圆点状，较小，分布在叶的中上部呈星状，叶子浓绿。当年新梢长 91.36 cm，粗 1.14 cm，叶数 85.5 片，接后 3 年每株新梢数 36.5 个，叶数 1 385 片，枝总长 13.49 m，叶面积指数 5.07。接后当年株产鲜叶 0.095 kg，2 年 0.54 kg，3 年 0.848 kg。短枝叶重占 36.4%，长枝占 63.61%。大叶、高产。黄酮含量 1.835%，内酯含量 0.085%，其中 GJ 0.01%、GC 0.030%、GA 0.024%、GB 0.002%、BB 0.013%。本系属于高产无性系（黄酮苷采用 754 分光光度法测，萜内酯采用 HPLC 法测据）。

'Fengchan-Y13'

Male. Mother tree in Shandong, the tree is 21.20 m tall and 65.00 cm in d.b.h, 100 years old, crown width are 7.50 m and 10.40 m, pole height 5.40 m, grows at the sea level 200 m.

Leaves deep green, fan-shaped, margin sinuolate, wedged at the base. One deep lobe, long 8.2 cm and wide 3.3 cm. Ocellus dense, punctate, distributed at the upper of leaf. Current shoots long 91.36 cm and with diameter 1.14 cm, each current shoots with 85.5 leaves, the leaves area index is 5.07. The each 3 years old grafting tree, with current shoots 36.5, 1 385 leaves, the total length of branches is 13.49 m.

The mean yields of fresh leaf of grafting trees are 0.095 kg (1st year), 0.54 kg (2nd year), and 0.848 kg (3rd year), which short branchlets yields 36.4% of it and long branchlets yields 63.61% of it. The current leaves, flavonoid glycosides content is 1.835% and lactone content is 0.085%, BB 0.013%, GJ 0.01%, GC 0.030%, GA 0.024% and GB 0.002%. Flavonoid glycoside determined by 754 Spectrophotometry and terpene lactone determined by HPLC (High Performance Liquid Chromatography).

11 '丰产 Y-7 号'

雄株，原产福建。60 年生，树高 14.00 m，胸径 0.40 m，生长旺盛，花粉量较大。属优良雄株。标准叶扇形，全缘，基部楔形，油胞极稀，呈斑状，点状分布在叶的中下部，叶浅绿色。每株新梢数每株 18 个，每株叶数 1 193 片，每株枝总长 10.92 m，冠幅 161.7 cm × 143 cm。当年新梢长 62.5 cm，粗 1.08 cm，每梢叶数 40 片。叶面积系数 3.43。接后当年株产鲜叶 0.116 kg，2 年 0.565 kg，3 年 0.925 kg。短枝叶重占 68.11%，长枝叶占 31.89%。总黄酮较高，达 2.06%（1 年），2 年生达 1.0%，3 年生达 0.76%。内酯总量 0.18%，其中 GJ 0.006%，GC 0.036%，GA 0.136%，GB 0.034 7%，BB 0.051 1%。该品种产量高，黄酮含量也较高（黄酮苷采用 754 分光光度法测，萜内酯采用 HPLC 法测据）。

'Fengchan-Y7'

Male. Mother tree in Fujian, the tree is 14.00 m tall and 40.00 cm in d.b.h, 60 years old, crown width are 1.617 m×1.430 m. Leaves light green, fan-shaped, entire margin, wedged at the base. Ocellus sparse, punctuate and nubbly, distributed at the lower of leaf. Current shoots long 62.5 cm with diameter 1.08 cm, each current shoots with 40 leaves, the leaves area index is 3.43. The each grafting tree, with current shoots 18, 1 193 leaves, the total length of branches is 10.92 m.

The mean yields of fresh leaf of grafting trees are 0.116 kg (1st year), 0.565 kg (2nd year), and 0.925 kg (3rd year), which short branchlets yields 68.11% of it and long branchlets yields 31.89% of it. The flavonoid glycosides content is 2.06% (1st year), 1.0% (2nd years) and 0.76% (3rd years). The mean value of lactone content is 0.18%, BB 0.051 1%, GJ 0.006%, GC 0.036%, GA 0.136% and GB 0.034 7%. Flavonoid glycoside determined by 754 Spectrophotometry and terpene lactone determined by HPLC (High Performance Liquid Chromatography).

3.4 银杏花粉用优良无性系

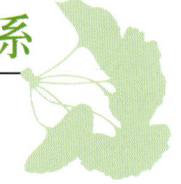

银杏花粉具有较高的营养价值与保健功能，被誉为"微型营养库"，花粉中含有人体必须的氨基酸、可利用矿物质元素和多种维生素，具有延缓皮肤衰老、防治肿瘤和心血管疾病等作用。同时银杏花粉中含有的黄酮类化合物，具有抗动脉硬化、降低胆固醇、防辐射等作用。

分别对3个银杏花粉用优良无性系进行简要介绍。

3.4 Ginkgo Clones for Pollen-producing

Ginkgo pollen has a higher nutritional value and health function, known as 'micro-nutrient bank'.There are essential amino acids, mineral elements and vitamins in pollen which can delaying senility of skin, preventing and treating tumor and cardiovascular diseases. Also, the Ginkgo flavonoids contained in pollen can be used in anti-atherosclerosis, lower cholesterol and anti-radiation.

Illustrated 3 Ginkgo clones for pollen-producing in brief.

1 '嵩优 1 号雄株'

该树位于河南洛阳嵩县，海拔 720 m，唐朝建庙时所植，树龄约 1 000 年。树高 17 m，主干高 7 m，胸径 1.78 m，冠幅 9 m×14 m，树冠较圆满。叶花丛节间短，数量多。1、2、3 年生枝平均雄球花长 3.18cm，粗 0.86 cm，花药 1 对，个数 72.78 个，每百个鲜雄球花重 25 g。每千克鲜雄球花 4 000 个，该雄株始花期 4 月 16 日，末花期 4 月 23 日。开花期 8 天，和雌株授粉期基本一致。

'Songyou-1（Xiongzhu）'

The tree is 17.00 m tall and 178 cm in d.b.h, main trunk is 7.00 m tall, about 1 000 years old, crown width are 9 m×14 m. The mean length of different ages staminate strobilus (one year old, two years old and three years old) is 3.18 cm with diameter 0.86 cm, one staminate strobilus with 72.78 anthers. Each one hundred fresh spica weight is 25 g, 1 000-grams-fresh male flowes has 4 000 male cones. The anthesis is 8 days form April 16th to April 23rd, the same as the pollination period of female.

图 3.216
河南洛阳嵩县'嵩优 1 号雄株'

Fig. 3.216
'Songyou-1（Xiongzhu）' in Songxian, Luoyang, Henan

2 '南林花 1'

雄株，来源于江苏泰兴，树龄 50 年。开花早，雄球花大，单个雄球花及单株花粉产量高，花粉萌发率高。嫁接后 4 年开始开花，雄球花，长度达 1.95 cm，直径 0.72 cm，单个小孢子叶球花粉量达到 8.04 mg；单株花粉产量高，5 年生花粉最高株产 0.25 kg，8 年生最高株产 0.86 kg，平均 0.64 kg，花粉萌发率高，达 87%。

'Nanlin Hua-1'

Male. The tree is 50 years old. Bloom early, male cone larger, with high pollen productivity and germinating rate. The grafting tree will come into blossom at 4 years old, male cone long 1.95 cm with diameter 0.72 cm, the mean weight of pollen of each male cone is about 8.04 mg; the highest pollen productivity of 5 years old and 8 years old grafting trees are 0.25 kg and 0.86 kg. The germinating rate is 87%.

图 3.217 江苏泰兴 '南林花 1'

Fig. 3.217 'Nanlin Hua-1' in Taixing, Jiangsu

3 '南林花 2'

雄株，来源于江苏泰兴，树龄 40 年。开花早，小孢子叶球大，单个小孢子叶球及单株花粉产量高，花粉萌发率高。嫁接后 4 年开始开花，小孢子叶球大，长度达 2.33 cm，直径 0.69 cm，单个小孢子叶球花粉量达到 8.44 mg；单株花粉产量高，5 年生花粉最高株产 0.31 kg，8 年生最高株产 0.94 kg，平均 0.71 kg，花粉萌发率高，达 85.5%。

'Nanlin Hua-2'

Male. The tree is 40 years old, male cone larger, with high pollen productivity and germinating rate. The grafting tree will come into blossom at 4 years old, male cone long 2.33 cm with diameter 0.69 cm, the mean weight of pollen of each male cone is about 8.44 mg; the highest pollen productivity of 5 years old and 8 years old grafting trees are 0.3l kg and 0.94 kg. The germinating rate is 85.5%.

图 3.218
江苏泰兴'南林花 2'

Fig. 3.218
'Nanlin Hua-2' in Taixing, Jiangsu

3.5 银杏材用优良无性系

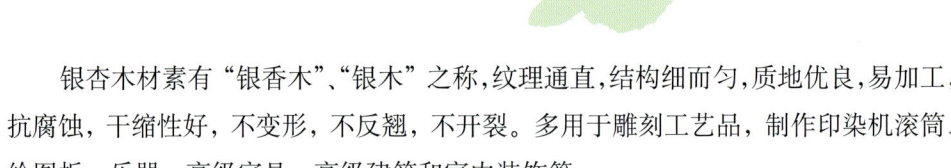

银杏木材素有"银香木"、"银木"之称,纹理通直,结构细而匀,质地优良,易加工,抗腐蚀,干缩性好,不变形,不反翘,不开裂。多用于雕刻工艺品,制作印染机滚筒、绘图板、乐器、高级家具、高级建筑和室内装饰等。

分别对3个银杏材用优良无性系进行简要介绍。

浙江诸暨古银杏群落
Ginkgo Trees in Zhuji, Zhejiang

3.5 Ginkgo Clones for Timber-producing

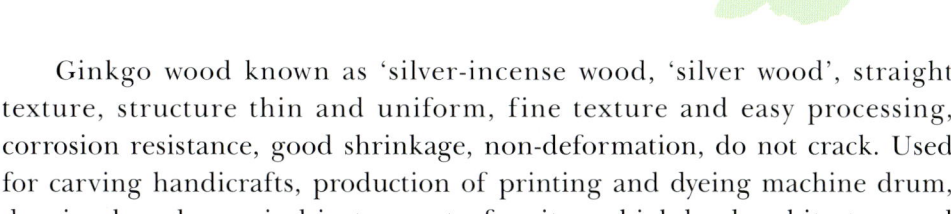

Ginkgo wood known as 'silver-incense wood, 'silver wood', straight texture, structure thin and uniform, fine texture and easy processing, corrosion resistance, good shrinkage, non-deformation, do not crack. Used for carving handicrafts, production of printing and dyeing machine drum, drawing boards, musical instruments, furniture, high-level architecture and interior decoration.

Illustrated 3 Ginkgo clones for timber-producing in brief.

1 '豫皖 9 号'

原株产于河南安阳，为速生用材型优树，雌株。胸径年生长量 1.6 cm，材积平均生长量 0.063 1 m³。其性状有待进一步观察。

'Yu Wan-9'

Female. The tree grows in Anyang, Henan, fast growing timber using tree. 1.6 cm in d.b.h growth and 0.063 1 m³ in average volume growth every year.

2 '直干银杏 S-31 号'

母树为实生，雌株，树龄为 400～500 年，高 16 m，胸径 1 m，枝条粗壮，生长旺盛。主干明显，直立生长，嫁接成活率 88.9%，每米长枝上短枝数 30 个，二次枝数 11 个，枝角小于 25.5°，成枝力大于 37%，适于材用。

'Zhigan-S31'

Female. The tree is 400~500 years old, up to 16 m height, 100 cm d.b.h, strong growth. Trunk obvious, vertical growth, graft survival rate is 88.9%, each branch with thirty branchlets, eleven branches (germination by repetition), branch angle less than 25.5°, branching ability more than 37%. Suitable for timber-forest.

3 '南林 B3'

南京林业大学选育，自由杂交种，速生，抗性强，树冠稀疏，透光率高，光合速率达 12 μmol·m⁻²·s⁻¹。年平均胸径生长 1.2 cm，年平均高生长超过 1.28 m。适于用作营造防护林、用材林和农田林网。

'Nanlin-B3'

Selected and bred by NJFU, natural hybrids, fast growing tree, strong resistance, sparse crown with high light transmittance, photosynthetic rate is up to 12 μmol·m⁻²·s⁻¹. 1.2 cm in d.b.h growth and 1.28 m in average growth every year. Suitable for shelter-forest, timber-forest and farmland shelterbelt network.

第4章 Chapter 4

银杏古树资源

Resources of Ancient Ginkgo Trees in China

第4章 银杏古树资源
Chapter 4　Resources of Ancient Ginkgo Trees in China

  银杏古树名木是大自然赋予人类的珍贵遗产，不仅是独特的自然和历史景观，而且有丰富的文化内涵，也是人类历史社会发展的佐证，对探索自然地理环境变迁、植物区系发生和发展规律，以及监测人类活动对自然环境的影响，都具有十分重要的意义，对研究人类历史文化也具有重要价值。

  中国是世界上银杏资源最丰富、分布最广泛的国家。据调查，浙江的天目山、湖北的大洪山、四川和湖北交界处的神农架、安徽和河南交界处的大别山有少量自然状态下的野生状态银杏群落，云南、贵州、广西等地的边远地区的山区也有自然零星生长的银杏，而银杏古树的分布则遍及全国绝大多数省份。据不完全统计，全国共有近30万棵银杏古树名木，分布于21个省（自治区）和4个直辖市的300个左右的县（市）区。如河北的抚宁，甘肃的康县，江苏的泰兴、邳州，安徽的金寨，山东的郯城，浙江的临安，贵州的福泉等县（市），均有银杏古树分布。

  中国部分银杏古树的分布明细见表4.1。

  C hina is rich in Ginkgo tree species. There are more than 300 000 old Ginkgo trees, mainly growing in Beijing, Zhejiang, Sichuan, Yunnan, Guizhou, Fujian and Shandong. We believe that the oldest Ginkgo tree grows in Fulai Mountain, Shandong, with 4.09 d.b.h and 26.4m height. But some people said that the Ginkgo tree growing in Fuquan, Guizhou may be the oldest one in China.

  Table 4.1 shows some information of 31 old Ginkgo trees in China.

银杏与恐龙
Ginkgo and Dinosaur

表 4.1 中国部分银杏古树的分布明细
Table 4.1 A brief list of ancient Ginkgo trees in China

	生长地点 Location	性别 Sex	树龄 Age (a)	树高 Height (m)	胸径 Diameter (m)	备注 Remark
1	安徽寿县报恩寺 Baoen Temple, Anhui		1 300	16.0	1.40	
2	北京市昌平区南口镇居庸关四桥子村 Juyongguan, Beijing	雌 female	1 200	25.0	2.50	
3	北京市门头沟区潭柘寺比庐殿前 Tanzhe Temple, Beijing	雄 male	1 000	30.0	2.90	清乾隆年间被封为 "帝王树" Emperor Tree
4	福建顺昌县大干镇 Shunchang, Fujian			40.0	2.00	
5	甘肃省陇南市徽县水阳乡沙嘴村 Huixian, Gansu	雌 female	1 700		2.00	
6	广东韶关市南雄油山 Nanxiong, Guangdong	雌 female	1 260			
7	广西壮族自治区恭城县三江乡新寨村 Gongcheng, Guangxi	雌 female	1 200	30.0	3.20	
8	贵州省福泉市黄丝镇邦乐村李家湾村 Fuquan, Guizhou	雄 male		40.1	4.70	
9	河北省三河市大掠马村 Daluema Village, Sanhe, Hebei	雌 female	1 300	30.0	2.94	
10	河北省遵化市侯家寨乡禅林寺院周围 Chanlin Temple, Zunhua, Hebei	雄 male	2 000	26.4	1.42	
11	河南省光山县大苏山净居寺 Jinju Temple, Guangshan, He'nan	雌 female	1 300	24.0	2.15	
12	河南王屋山 Wangwu Mountain, He'nan		2 000	45.7	3.00	
13	湖南省洞口县那溪镇宝瑶村 Dongkou, Hu'nan	雌 female	1 500	48.0	4.60	
14	江苏省无锡市灵山祥福寺 Xiangfu Temple, Wuxi, Jiangsu	雌 female	1 400			
15	江苏省宜兴市周墅乡双乔村 Yixing, Jiangsu		1 000	17.0	1.09	孙权之母手植银杏 Planted by Sun Quan's mother
16	江西省九江市庐山区莲花镇宝积庵 Baoji Hut, Jiujiang, Jiangxi		1 000	20.5	3.02	
17	江西省永修县云居山真如寺 Zhenru Temple, Yongxiu, Jiangxi		1 400	30.0	2.23	

续表

	生长地点 Location	性别 Sex	树龄 Age (a)	树高 Height (m)	胸径 Diameter (m)	备注 Remark
18	山东曲阜孔庙 Confucian temple, Qufu, Shandong	雌 female	1 000			
19	山东曲阜孔庙 Confucian temple, Qufu, Shandong	雄 male	1 000			
20	山东省莒县浮来山定林寺 Dinglin Temple, Ju county, Shandong	雌 female	3 300	26.4	4.09	
21	山东省沂源县织女洞三清殿 Sanqing Hall, Yiyuan, Shandong	雌 female		24.5	1.02	叶籽银杏 Ye Zi Yinxing
22	山东泗水县石莱镇白马寺 Baima Temple, Sishui, Shandong		2 000	21.0	2.87	
23	陕西汉中巴县赤南乡庄龙坪村白果坪 Hanzhong, Shanxi	雌 female	1 200	35.0	3.00	
24	陕西省略阳县青泥河乡琵琶寺 Pipa Temple, Lueyang, Shanxi	雌 female	1 300	28.0	2.29	李白手植 Planted by Li Bai
25	陕西周至县 Zhouzhi, Shanxi	雄 male	2 600	24.0	3.00	老子手植银杏 Planted by Lao Zi
26	上海 Shanghai		1 400	24.5	2.00	
27	四川都江堰 Dujiang Yan, Sichuan	雌 female	1 700	16.3	1.60	张松手植银杏 Planted by Zhang Song
28	四川省都江堰市青城山天师洞祖师殿 Zushi Hall, Dujiang Weir, Sichuan	雄 male	1 800	29.5	2.42	
29	云南省腾冲县界头乡白果村 Tengchong, Yunnan	雌 female	1 000	30.0	2.30	
30	浙江省西天目山禅源寺开山老殿下 Chanyuan Temple, Lin'an, Zhejiang	雌 female	2 000	23.5	1.12	长势良好，与其他20多株连生，有"五世同堂"之称 Main stems with different ages
31	重庆市南川市半河乡大河村 Nanchuan, Chongqing	雄 male	2 500	26.0	3.00	

银杏古树资源
Chapter 4　Resources of Ancient Ginkgo Trees in China

图 4.1　北京潭柘寺帝王树
Fig. 4.1　The Ancient Ginkgo in Tanzhe Temple, Beijing

图 4.2 甘肃陇南沙嘴村古银杏
Fig. 4.2 The Ancient Ginkgo in Jianzui of Longnan, Gansu

图 4.3 贵州福泉大银杏
Fig. 4.3 The Ginkgo in Fuquan, Guizhou

图 4.4 河北禅林寺古银杏（龙种）

Fig. 4.4　The Ancient Ginkgo of Long Zhong in Chanlin Temple, Hebei

图 4.5 河北大掠马
Fig. 4.5 The Ancient Ginkgo of Da Luema in Hebei

图 4.6　河南王屋山古银杏

Fig. 4.6　The Ancient Ginkgo in Wangwu Mountain, Henan

银杏古树资源

Chapter 4　Resources of Ancient Ginkgo Trees in China

图 4.7　山东莒县浮来山定林寺
Fig. 4.7　The Ancient Ginkgo in Dinglin Temple, Fulai Mountain, Shandong

图 4.8 山东曲阜孔庙宋代古银杏
Fig. 4.8 The Ancient Ginkgo of Song Dynasty of Kong Temple in Qufu, Shandong

Chapter 4　Resources of Ancient Ginkgo Trees in China

图 4.9　山东泗水白马寺
Fig. 4.9　The Ancient Ginkgo in White Horse Temple in Sishui, Shandong

图 4.10　山东沂源县叶籽银杏树
Fig. 4.10　Ye Zi Yinxing in Yiyuan, Shandong

图 4.11 陕西汉中银杏
Fig. 4.11 The Ancient Ginkgo in Hanzhong, Shaanxi

图 4.12　陕西周至县老子手植银杏
Fig. 4.12　The Ginkgo Planted by Lao Tzu in Zhouzhi, Shaanxi

Chapter 4　Resources of Ancient Ginkgo Trees in China

图 4.11　陕西汉中银杏
Fig. 4.11　The Ancient Ginkgo in Hanzhong, Shaanxi

第4章 银杏古树资源
Chapter 4　Resources of Ancient Ginkgo Trees in China

图 4.12　陕西周至县老子手植银杏
Fig. 4.12　The Ginkgo Planted by Lao Tzu in Zhouzhi, Shaanxi

Chapter 4　Resources of Ancient Ginkgo Trees in China

图 4.13　上海银杏王
Fig. 4.13　The Ginkgo King in Shanghai

第4章 银杏古树资源
Chapter 4　Resources of Ancient Ginkgo Trees in China

图 4.14　四川都江堰张松银杏北侧
Fig. 4.14　The Ginkgo Planted by Zhang Song in Dujiang Yan, Sichuan

图 4.15 云南滕冲千年银杏姐妹树
Fig. 4.15 Millennium Ginkgo of Two Sisters in Tengchong, Yunnan

图 4.16 浙江天目山"五代同堂"
Fig. 4.16 Five-generation Ginkgo on Tianmu Montain, Zhejiang

参考文献
References

曹福亮. 2000. 银杏资源培育与高效利用 [M]. 北京：中国林业出版社.

曹福亮. 2003. 中国银杏 [M]. 南京：江苏科学技术出版社.

曹福亮. 2007. 银杏（画册）[M]. 北京：中国林业出版社.

曹福亮. 2007. 中国银杏志 [M]. 北京：中国林业出版社.

曹福亮，徐立安，丁翠柏，等. 1996. 银杏叶用林优良品种和优良单株选择 [J]. 江苏林业科技，23(1): 12-14.

曹福亮，黄敏仁，桂仁意. 2005. 银杏主要栽培品种遗传多样性分析 [J]. 南京林业大学学报（自然科学版），29(6): 1-6.

曹福亮，桂仁意，张往祥，等. 2007. 银杏主要栽培品种种子性状变异及聚类分析：银杏资源培育及高效利用 [M]. 北京：科学技术文献出版社：134-146.

曹福亮，王国霞，李广平，等. 2008. 银杏 ISSR-PCR 扩增反应体系的优化 [J]. 浙江林学院学报，25(2): 186-190.

曹福亮，花喆斌，汪贵斌，等. 2008. 野生银杏资源群体遗传多样性的 RAPD 分析 [J]. 浙江林学院学报，25(1): 22-27.

陈俊愉. 2001. 中国花卉品种分类学 [M]. 北京：中国林业出版社.

郝明灼，曹福亮，张往祥，等. 2006. 银杏不同雄株花粉外观形态的变异 [J]. 林业科技开发，20(5): 53-55.

莫昭展，曹福亮，符韵林. 2007. 银杏雄株种质资源遗传多样性研究 [J]. 安徽农业科学，35(26): 8130, 8131, 8133.

向其柏，臧德奎，孙卫邦，等. 2006. 国际栽培植物命名法规 [M]. 第七版. 北京：中国林业出版社.

Boland T, et al. 2002. Michigan gardener's guide [M]. Cool Springs Press.

Brickell C D, Baum B R, et al. 2004. International code of nomenclature for cultivated plants [M]. ISHS. Acta Horticulturae: 647.

Elsabagh S, et al. 2005. Differential cognitive effects of *Ginkgo biloba* after acute and chronic treatment in healthy young volunteers [J]. Psychopharmacology (Berl), 179 (2): 437-446.

Ernst E, Canter P H, Coon J T. 2005. Does *Ginkgo biloba* increase the risk of bleeding? A systematic review of case reports [J]. Perfusion, 18: 52-56.

Fu Liguo, et al. 1999. *Ginkgo biloba* [J]. Flora of China, 4, Beijing: Science Press, St. Louis: Missouri Botanical Garden Press: 8.

References

Holt B F, Rothwell G W. 1997. Is *Ginkgo biloba* (Ginkgoaceae) really an oviparous plant [J]. American Journal of Botany, 84: 6.

Lewington A, Parker E. 1997. Ancient trees [M]. London: Collins & Brown Ltd.:183.

Mahadevan S, Park Y. 2008. Multifaceted therapeutic benefits of *Ginkgo biloba* L.: Chemistry, Efficacy, Safety, and Uses [J]. J. Food Sci.,1 (14-9): 73.

Mo Zhaozhan, Cao Fuliang, Wang Guibin, et al. 2006. Analysis of variation of seed traits of Ginkgo [J]. Journal of Hebei Forestry Science and Technology, (4):1-5.

Rehder A. 1940. Bibliography of cultivated trees and shrubs hardy in Northern America [M]. 2nd ed. New York: The MaCmillan Company.

Simpson D P. 1979. Cassell's Latin Dictionary [M]. 5th ed. London: Cassell Ltd.: 883.

Smith P F, Maclennan K, Darlington C L. 1996. The Neuroprotective properties of the *Ginkgo biloba* leaf: A review of the possible relationship to platelet-activating factor (PAF) [J]. Journal of Ethnopharmacology, 50 (3): 131-139.

Taylor Thomas N, Edith L Taylor. 1993. The biology and evolution of fossil plants [J]. Englewood Cliffs: 138,197.

Trehane P, Brickell C D, Baum B R, et al. 1995. International code of nomenclature for cultivated plants [M].Wimborne, UK.

Witkam L and Ramzan I. 2004. *Ginkgo biloba* in the treatment of alzheimer's disease: A miracle cure? [J]. From Cell to Society.

Xuemin Jiang, et al. 2005.Effect of ginkgo and ginger on the pharmacokinetics and pharmacodynamics of warfarin in healthy subjects [J]. British Journal of Clinical Pharmacology, 59 (4): 425-432.

Zhou Zhiyan, Zheng Shaolin. 2003. Palaeobiology: The missing link in Ginkgo evolution [J]. Nature, 423 (423): 821-822.

索 引

1. 品种中文名索引

'安吉 F4'　34,46
'安陆 1 号'　106,134,135
'安陆 64 号'　120
'安陆 A11'　34,39
'安陆 A14 号'　136,142
'安陆 A3-1 号'　136,148
'安陆大白果'　134
'扁佛指'　50,76,77
'曹 1 号'　78,85
'曹 2 号'　78,84
'长白果'　32
'长糯白果'　50,58,59
'长兴 1 号'　34,35
'长兴 2 号'　79,101
'长兴 3 号'　34,37
'长兴 4 号'　34,38
'长兴 5 号'　78,90
'长兴 F13'　34,43
长子品种群　20,22,24,34
'垂枝银杏'　184,186
'大金果'　112
'大金坠'　22,32,33
'大龙眼'　154
'大梅核'　128
'大圆头'　166
'大圆珠'　166
'大圆子'　152,166,167
'大钻头'　24
'道真 5 号'　176,177
'道真 7 号'　176,178
'东山 F15'　34,47
'洞庭佛手 1 号'　79,97
'洞庭皇'　50,62,63
'23 号大梅核'　132
'费尔蒙特'　184,194
'丰产 Y-3 号'　204
'丰产 Y-6 号'　204
'丰产 Y-7 号'　205
'丰产 Y-8 号'　203
'佛指'　60
佛指品种群　20,50,52,78
'橄榄佛手'　24
'橄榄果'　22,24,25
'港上 303 号'　79,95

'港上 309 号'　136,150
'港上 501 号'　79,102
'港西 2 号'　78,88
'高优 Y-2 号'　202
'古银杏'　79,104
'观音皇'　106,120,121
'贵州长白果'　50,66,67
'桂 047 号'　152,158,159
'桂 048 号'　152,160,161
'桂 049 号'　106,118,119
'桂林 6 号'　176,180
'桂林 8 号'　136,149
'桂林 9 号'　78,80
'海洋皇'　106,122,123
'海洋王'　122
'华口大白果'　50,72,73
'黄皮果'　50,68,69
'黄酮 F-1 号'　198
'黄酮 F-2 号'　198
'黄酮 F-3 号'　199
'家佛指'　50,60,61
'金兵普林斯顿'　184,194
'金秋-1'　184,193
'金球'　184,195
'金坠 13 号'　50,70,71
'金坠 1 号'　74
'京山 A23 号'　176,181
'京山 A25 号'　78,82
'九甫长籽'　22,30,31
'魁金'　106,112,113
'魁铃'　106,114,115
'李子果'　106,110,111
'灵川 F9'　34,42
'龙眼'　152,164,165
'马铃 3 号'　114
'马铃 5 号'　106,126,127
马铃亚品种群　20,105,106
'梅核'　106,128,129
梅核亚品种群　20,105,106
'棉花果'　106,130,131
'内酯 GB-5 号'　201
'内酯 T-5 号'　200

'内酯 T-6 号'　200
'南林 B3'　212
'南林花 1'　208
'南林花 2'　209
'邳县大马铃'　108
'葡萄果'　152,162,163
'七星果'　50,52,53
'青皮果'　50,54,55
'胜利 102 号'　34,41
'嵩优 1 号雄株'　207
'苏农佛手'　79,98
'塔状银杏'　184,188
'泰兴 1 号'　78,91
'泰兴 2 号'　79,103
'泰兴 3 号'　78,89
'泰兴 4 号'　34,36
'泰兴大白果'　60
'郯 306 号'　64
'郯城 107 号'　156
'郯城 16 号'　136,147
'郯城 207 号'　136,139
'郯城 231 号'　79,94
'郯城 322 号'　136,144
'郯城 5 号'　126
'郯城 9 号'　136,143
'郯丰'　152,156,157
'郯魁'　50,64,65
'郯马 1 号'　136,137
'郯新'　78,83
'藤九郎'　176,182
'天目长籽'　22,28,29
'铁富马铃 3 号'　124
'铁马 1 号'　136,141
'桐子果'　152,172,173
'筒叶银杏'　184,192
'团峰'　152,154,155
'万年金'　184,189
'五月田野'　184,194
'狭叶银杏'　184,191
'小梅核'　128
'小圆头'　170
'小圆珠'　170
'小圆子'　152,170,171
'新村 16 号'　136,140

'新村 18 号'　136,146
'新村 202 号'　78,86
'新村 203 号'　34,49
'新村 210 号'　79,96
'新村 222 号'　34,45
'新村 231 号'　34,40
'新村 401 号'　79,99
'新村 402 号'　34,44
'新村 5 号'　126
'新村 9 号'　78,92
'新宇'　50,74,75
'鸭尾股银杏'　56
'鸭尾银杏'　50,56,57
'亚甜'　106,108,109
'延安 1 号'　176,179
'叶籽银杏'　78,81
'宇香'　106,124,125
'玉蝴蝶'　184,190
'豫皖 9 号'　212
圆子品种群　20,151,152,154,176
'圆底佛手'　116
'圆底果'　106,116,117
'圆铃'　152,168,169
'圆铃 6 号'　154
'圆枣佛手'　22,26,27
'早实梅核'　106,132,133
'枣子佛手'　26
'枣子果'　26
'展冠银杏'　184,187
'掌状银杏'　184,191
'正安 1 号'　79,100
'正安 3 号'　136,145
'正安 5 号'　136,138
'直干银杏 S-31 号'　212
中子品种群　20,105,106,108,136
'重坊 106 号'　78,87
'重坊 111 号'　78,93
'重坊 176 号'　34,48
'皱皮果'　152,174,175

2. 品种拉丁名索引

'Anji-F4'　　34, 46
'Anlu -A11'　　34, 39
'Anlu Da Baiguo'　　134
'Anlu-1'　　107, 134, 135
'Anlu-64'　　120
'Anlu-A14'　　136, 142
'Anlu-A3-1'　　136, 148
'Autumn Gold'　　193
'Bian Fozhi'　　51, 76, 77
'Cao-1'　　78, 85
'Cao-2'　　78, 84
'Chang Baiguo'　　32
'Changnuo Baiguo'　　51, 58, 59
'Changxing-1'　　34, 35
'Changxing-2'　　79, 101
'Changxing-3'　　34, 37
'Changxing-4'　　34, 38
'Changxing-5'　　78, 90
'Changxing-F13'　　34, 43
Changzi Group　　21, 23, 24, 34,
'Chongfang-106'　　78, 87
'Chongfang-111'　　78, 93
'Chongfang-176'　　34, 48
'Chuizhi Ginkgo'　　184, 186
'Da Jin Guo'　　112
'Da Jinzhui'　　23, 32, 33
'Da Longyan'　　154
'Da Meihe 23'　　132
'Da Meihe'　　128
'Da yuantou'　　166
'Da Yuanzhu'　　166
'Da Yuanzi'　　153, 166, 167
'Da Zuantou'　　24
'Daozhen-5'　　176, 177
'Daozhen-7'　　176, 178
'Dissected Ginkgo'　　191
'Dongshan-F15'　　34, 47
'Dongting Foshou-1'　　79, 97
'Dongting Huang'　　51, 62, 63
'Er Jie Tou'　　112

'Fairmount'　　184, 194
'Fastigiata'　　194
'Feiermengte'　　184, 194
'Fengchan-Y13'　　204
'Fengchan-Y6'　　204
'Fengchan-Y7'　　205
'Fengchan-Y8'　　203
Fozhi Group　　21, 51, 52, 78
'Fozhi'　　60
'Gangshang-303'　　79, 95
'Gangshang-309'　　136, 150
'Gangshang-501'　　79, 102
'Gangxi-2'　　78, 88
'Ganlan Foshou'　　24
'Ganlan Guo'　　23, 24, 25
'Gaoyou-Y2'　　202
'Golden Globe'　　195
'Gu Yinxing'　　79, 104, 105, 151
'Guanyin Huang'　　107, 120, 121
'Gui-047'　　153, 158, 159
'Gui-048'　　153, 160, 161
'Gui-049'　　107, 118, 119
'Guilin-6'　　176, 180
'Guilin-8'　　136, 149
'Guilin-9'　　78, 80
'Guizhou Chang Baiguo'　　51, 66, 67
'Haiyang Huang'　　107, 122, 123
'Haiyang Wang'　　122
'Horizontalis'　　187
'Huakou Da Baiguo'　　51, 72, 73
'Huangpi Guo'　　51, 68, 69
'Huangtong-F1'　　198
'Huangtong-F2'　　199
'Huangtong-F3'　　199
'Jade Butterfly'　　190
'Jia Fozhi'　　51, 60, 61
'Jin Qiu'　　184, 195
'Jin Qiu-1'　　184, 193
'Jinbing Pulinsidun'　　184, 194
'Jingshan-A23'　　176, 181

'Jingshan-A25'　　78, 82
'Jinzhui-1'　　74
'Jinzhui-13'　　51, 70, 71
'Jiufu Changzi'　　23, 30, 31
'Kui Jin'　　107, 112, 113
'Kui Ling'　　107, 114, 115
'Lingchuan-F9'　　34, 42
'Lizi Guo'　　107, 110, 111
'Longyan'　　153, 164, 165
'Maling-3'　　114
'Maling-5'　　107, 126, 127
'Mariken'　　188
'Mayfield'　　184, 194
'Meihe'　　107, 128, 129
'Mianhua Guo'　　107, 130, 131
'Nanlin Hua-1'　　208
'Nanlin Hua-2'　　209
'Nanlin-B3'　　212
'Neizhi-GB5'　　202
'Neizhi-T5'　　200
'Neizhi-T6'　　201
'Pendula'　　186
'Pixian Da Maling'　　108
'Princeton Sentry'　　184, 194
'Putao Guo'　　153, 162, 163
'Qingpi Guo'　　51, 54, 55
'Qixing Guo'　　51, 52, 53
'Saratoga'　　191
'Shengli-102'　　34, 41
'Songyou-1（Xiongzhu）'　　207
Maling Sub-group　　21, 105, 107
Meihe Sub-group　　21, 105, 107
'Sunong Foshou'　　79, 98
'Taixing Dabaiguo'　　60
'Taixing-1'　　78, 91
'Taixing-2'　　79, 103
'Taixing-3'　　78, 89
'Taixing-4'　　34, 36
'Tan Feng'　　153, 156, 157
'Tan Kui'　　51, 64, 65

'Tan Ma-1'   136, 137
'Tan Xin'   78, 83
'Tan-360'   64
'Tancheng-107'   156
'Tancheng-16'   136, 147
'Tancheng-207'   136, 139
'Tancheng-231'   79, 94
'Tancheng-322'   136, 144
'Tancheng-5'   126
'Tancheng-9'   136, 143
'Tazhuang Ginkgo'   184, 188
'Teng Jiulang'   176, 182
'Tianmu Changzi'   23, 28, 29
'Tie Ma-1'   136, 141
'Tiefu Maling-3'   124
'Tongye Ginkgo'   184, 192
'Tongzi Guo'   153, 172, 173
'Tuan Feng'   153, 154, 155
'Tubifolia'   192
'Tubiformis'   192
'Wannian Jin'   184, 189
'Wuyue Tianye'   184, 194

'Xiao Meihe'   128
'Xiao Yuantou'   170
'Xiao Yuanzhu'   170
'Xiao Yuanzi'   153, 170, 171
'Xiaye Ginkgo'   184, 191
'Xin Yu'   51, 74, 75
'Xincun-16'   136, 140
'Xincun-18'   136, 146
'Xincun-202'   78, 86
'Xincun-203'   34, 49
'Xincun-210'   79, 96
'Xincun-222'   34, 45
'Xincun-231'   34, 40
'Xincun-401'   79, 99
'Xincun-402'   34, 44
'Xincun-5'   126
'Xincun-9'   78, 92
'Ya Tian'   107, 108, 109
'Yanan-1'   176, 179
'Yapigu Yinxing'   56
'Yawei Yinxing'   51, 56, 57
'Ye Zi Yinxing'   78, 81

'Yu Hudie'   184, 190
'Yu Wan-9'   212
'Yuandi Foshou'   116
'Yuandi Guo'   107, 116, 117
'Yuanling'   153, 168, 169
'Yuanling-6'   154
'Yuanzao Foshou'   23, 26, 27
Yuanzi Group   21, 151, 153, 154, 176
'Yuxiang'   107, 124, 125
'Zaoshi Meihe'   107, 132, 133
'Zaozi Foshou'   26
'Zaozi Guo'   26
'Zhanguan Ginkgo'   184, 187
'Zhangzhuang Ginkgo'   184, 191
'Zhengan-1'   79, 100
'Zhengan-3'   136, 145
'Zhengan-5'   136, 138
'Zhigan-S31'   212
Zhongzi Group   21, 105, 107, 108, 136
'Zhoupi Guo'   153, 174, 175